SEASHORE PLANTS
OF
SOUTH FLORIDA
AND THE
CARIBBEAN

A GUIDE TO IDENTIFICATION
AND PROPAGATION OF
XERISCAPE PLANTS

BY DAVID W. NELLIS

PINEAPPLE PRESS, INC.
SARASOTA, FLORIDA

SEASHORE PLANTS
OF
SOUTH FLORIDA
AND THE
CARIBBEAN

A GUIDE TO IDENTIFICATION
AND PROPAGATION OF
XERISCAPE PLANTS

PINEAPPLE PRESS, INC.
P.O. Drawer 16008
Southside Station
Sarasota, Florida 34239

Library of Congress Cataloging-in-Publication Data

Nellis, David W.
Seashore Plants of South Florida and the Caribbean/David W. Nellis-
1st ed.
    p. cm.
Includes bibliographical references and index.
ISBN 1-56164-026-3: Hb ISBN 1-56164-056-5 Pb
1. Seashore plants—Florida. 2. Seashore plants—Caribbean Area. 3. Seashore plants—Florida—Identification. 4. Seashore plants—Caribbean Area—Identification 5. Seashore plants—Florida—Pictorial works. 6. Seashore plants—Caribbean Area-Pictorial works. I. Title
QK 154.N43   1994
581.9759—dc20

93-40713
CIP

First Edition
10 9 8 7 6 5 4 3 2 1

Design & Composition
ROBERT FLEURY

PREFACE

This book is intended for use by those who feel a great affinity for that area of our planet where land and water meet. Casual strollers on the beach will find their outings more enjoyable if they know about the plants they meet. Homeowners and resort planners will find that selecting salt-tolerant plants for waterfront plantings will yield a more successful landscape with lower maintenance cost than one developed using standard ornamental plants. Owners of dredge-filled land will find that the use of salt-tolerant plants will eliminate many of the mysterious difficulties they have been experiencing with producing an attractive landscape.

The information presented here is derived from personal experience combined with information supplied by horticulturists and hundreds of published sources. Some of the most useful publications are listed after the index. All works of this nature build on the knowledge of our predecessors, which is often free for the asking at major libraries throughout the country. The libraries found to be particularly rich sources of information for the preparation of this book have been the Harvard University Botanical Libraries, the New York Botanical Garden Library, National Agricultural Library, Yale University Library, University of Georgia Science Library, Fairchild Tropical Garden Library (Montgomery Foundation Library) and the Institute of Tropical Forestry Library. In every case I have found the librarians to be knowledgeable and enthusiastic guides to the riches of the collections under their care. If you wish to know more about plants, I heartily recommend that you visit these vast underutilized resources.

ACKNOWLEDGMENTS

Many people, from librarians to horticulturists to one who has a carefully tended ornate flowerbed near the backyard vegetable garden, have supported my enthusiasm and assisted with the preparation of this book. Some of those who have spent the most time in providing information are Kathy Boone, Robert Campbell, Susan Curtiss, Cynthia Flack, Eleanor Gibney, Margaret Hayes, Julia Morton, Gizelle Reyes, Dee Serage, Donna Sprunt, and Toni Thomas. I have been honored by some of the grand old scholars of Caribbean botany taking time from their demanding schedules to review this work. F. Raymond Fosberg, George R. Proctor and Roy O. Woodbury all examined the text and provided useful comments. As with experts in many fields, they did not completely agree with each other or with published sources. I accept full responsibility for the final decision on the resolution of sticky problems of taxonomy and nomenclature. Roy Woodbury generously reviewed the slides to see that none were too far afield in the depiction of the correct plant.

TABLE OF CONTENTS

THE PLANTS

Dildo Cactus *Pilosocereus royenii*

INTRODUCTION

The seashore is a hostile environment for plants. The characteristics of a beach make fresh water a scarcity.

- Rain that falls drains away rapidly through the sand.
- Moisture is swiftly evaporated by the wind which is so prevalent on the seacoast.
- Salt in the form of spray or seeping seawater reduces the osmotic absorption of water.

Plants suitable for seashore planting thus have many of the characteristics of desert plants. Leaves have waxy surfaces to control water loss or are covered with fine hairs to reduce the absorption of heat. Root structures may be shallow and widespread to take advantage of brief showers, or a long and aggressive tap root may be present to reach deeply for moisture. The roots of salt-tolerant plants may have special physiological capabilities for rejecting salt. The leaves and stems of seashore plants may possess glands for expelling excess salt from the tissue. Reduction of leaf size and number is a common strategy to reduce water demand for plants growing on salty soils. As in the desert, seashore plants tend to be woody and slow growing due to limited resources, or fleshy, succulent and fast growing to take advantage of brief periods of favorable conditions.

The seashore is very dynamic over the scale of years. Sedimentation and erosion add and subtract from the shoreline, and periodic storm pounding renders the strand an untenable habitat for many plants, yet many of these processes are themselves greatly influenced by plants. Mangroves and marsh grass contribute significantly to sedimentation and land building. Figs and other rock-dwelling plants actively crack and dismantle rocky shorelines.

THE ECOLOGY OF THE SALTY ENVIRONMENT

At best, plants may tolerate, strategically avoid or otherwise cope with salinity, but usually they grow better under conditions of lower salinity. Even the most salt-loving plants do not thrive under highly saline conditions.

The amount of salt in the soil which a plant can tolerate depends on the species of plant along with the texture and water-holding capacity of the soil. Soil moisture near the sea often varies tremendously with only inches of difference in elevation or a few feet in distance from standing water.

The amount of salt spray and salt crystals in the air is inversely related to distance from the sea. Plants which grow on the very edge of the sea have a tolerance of salt in air and water which gives them an absolute competitive advantage. They usually cannot compete with other plants on inland sites with less salt. This tier of plants is typically composed of grasses or low-growing forbs and is quite mobile with respect to changing shorelines. Behind and inland from the frontline plants is a group of plants which are more shrubby in nature and are not quite as tolerant of wind and salt. Further inland and generally delineating a more stable shoreline are salt- and wind-tolerant trees. These coastal forests provide protection for the natural forest typical of the physiography of the region. We often fail to recognize this sequence of subtle intergraded protective succession until one of the tiers is removed. The resulting domino effect of each tier succumbing in sequence to salt and wind can be dramatic.

SELECTED PLANTS

The plants chosen for treatment in this book are all salt tolerant and are either ubiquitous in distribution or show considerable potential for intentional planting. In island archipelagoes it is common for one genus of plants to be represented by different species on different islands. In the Bahamas and Caribbean this is certainly true — many species are sprinkled throughout the range covered by this book. I have generally used the most widespread or salt tolerant of the species within our area as a representative of each genus. Most of the plants discussed are native or have become naturalized by reproducing successfully in the wild. A few are included by virtue of their eminent suitability for seashore landscaping.

PLANT NAMES

The scientific names of organisms are intended to provide a single universal system of reference. I have used these names for the plants in this account to provide a basic reference for those wishing to go further in the botanical literature. To somewhat dispel the mystique of these Latin names I have included the translation or origin of each name. While some are named to memorialize accomplished botanists, many others help to describe the plant itself.

As a result of continuous study and research, the classification of plants is revised over time, with new discoveries resulting in changes in the "accepted" scientific name. Thus over the course of history the same plant may be referenced under several different names. To facilitate the quest of those who would like to gather further information, I have included other recently used names in the notes at the end of each plant account.

The word mangrove is used with three different plants in this book. This word is applied to trees of several families and over 60 species worldwide and describes a series of shared characteristics. All have evolved a competitive advantage over other woody plants by developing physiological characteristics which allow them to thrive in seawater-moistened and often anaerobic soil. Most species develop prop roots for support in the soft environment and pneumatophores or other root structures to provide access to above-ground oxygen. The seeds of mangroves usually germinate while still attached to the tree and maintain their viability while being widely dispersed by the tides.

MEDICINAL USES

The medicinal uses reported herein are not recommendations but rather an indication of how the plants have been used in home remedies and folklore. In some instances, such as the use of aloe for treating burns, the use has been subsequently confirmed as effective by modern medicine. In other instances folklore has connected the appearance of a plant to an affliction to be treated. Using the scorpioid flower clusters of seaside heliotrope to treat scorpion sting and the frequent application of teas from red flowers to treat inflamed eyes are two of the many "like treats like" approaches to herbal medicine.

EDIBILITY

Many of the wild plants described in this book have minute amounts of potentially toxic compounds but have been consumed as food for hundreds of years. Most plants have developed toxic substances in their tissues to repel herbivores or at least limit their consumption of foliage. Many of these repellent substances can be tolerated by mammals if consumed in small amounts. The nip of horseradish, the bitterness of green olives, and the pungence of garlic are all examples of the delight of toxins in moderation. Chronic excessive exposure to many of these toxins can have adverse effects on the body. Much of the world's population regularly consumes teas which contain greater or lesser amounts of tannin. These tannins in the dust breathed by leatherworkers and woodworkers produce elevated levels of nasal and sinus cancer. Moderation, both in quantity consumed per serving and in regularity of consumption of any particular plant, is the best way to avoid subtle untoward side effects.

PLANT CHEMISTRY

The plant kingdom is a virtual cornucopia of chemical compounds. Many of these compounds are generated as plants build increasingly complex molecules starting from the simple sugars produced by sunlight and photosynthesis. The addition of nitrogen allows the construction of amino acids which in turn are the building blocks of proteins. As with any chemical factory, all these syntheses produce many secondary compounds, some of which are of no known use to the plant but others of which are toxic and may serve as defensive chemicals. Plants which incorporate chemicals that make them less desirable to consumers tend to thrive and become more abundant. The defensive chemicals vary extremely in their mode of action and chemical makeup.

Alkaloids

One of the more common groups of chemicals is called alkaloids. When used with reference to plant chemicals, this term is not a strict chemical designation but rather an inclusive term describing a wide array of nitrogen-bearing basic substances with ringlike structures, most of which possess the following characteristics:
- Complex chemical compounds produced by the plants themselves.

- Biochemically derived from various amino acids.
- Usually bitter tasting.
- Alkaline pH.
- Poorly soluble in water but more soluble in organic solvents.
- May be precipitated from solution by certain reagents.
- Provide a characteristic color reaction with certain reagents.
- Susceptible to destruction by heat.
- Degraded by exposure to light and air.
- Pharmacologically active in small amounts to various organs and tissues of animals.

Many of the alkaloids influence the nervous system in various ways: morphine acts as a sedative and narcotic, cocaine is a general stimulant, strychnine is a lethal convulsant and others produce hallucinations. We purchase plants in the grocery store for their alkaloid content: caffeine in coffee, theobromine in chocolate and nicotine in tobacco products.

Many pharmaceuticals are alkaloids that are extracted from plants, such as quinine for the treatment of malaria and the curare alkaloids for producing muscle relaxation in certain surgical procedures. (Curare used on blowgun darts by South American natives induces death by relaxation of the respiratory muscles). Many other pharmaceutical chemicals have been synthesized or modified to approximate the action of natural compounds.

Glycosides

The glycosides vary greatly in structure but all are two- or three-part molecules of which one part is a simple sugar. As intact molecules they are generally relatively inactive. When digestive processes separate the parts of the molecule, the released nonsugar part frequently has significant physiological influence.

In the cyanogenic glycosides, hydrocyanic acid (cyanide) is released. The cyanogenic glycoside amygdalin is extracted from apricot pits and marketed outside the U.S. as the controversial cancer drug laetrile.

In cardiac glycosides a lactone ring is connected to a steroid nucleus, yielding a drug which increases the force and decreases the rate of the heartbeat. Digitalis, commonly prescribed to treat heart problems, is a cardiac glycoside derived from the foxglove plant. The much more potent cardiac glycoside in the oleander has been responsible for many deaths in unsuspecting consumers.

The saponin glycosides have a steroid skeleton without the lactone ring and exert their physiological influence by causing severe gastric irritation. When the sapogenins are absorbed into the bloodstream they may cause rupture of the red blood cells due to interaction with the cholesterol in the cell membrane. This cholesterol interaction is being investigated in some of the more benign saponins in beans and grains as a method of reducing high levels of blood cholesterol.

The coumarin glycosides contain a phenolic-type molecule and are primarily noted as preventing clotting in mammalian blood. This factor is used in human medications and in an extreme example as the rat poison Warfarin, which causes lethal internal hemorrhage.

The anthraquinone glycosides are usually irritating to the intestinal tract and cathartic in effect. These are the active compounds in the traditional commercial plant cathartics aloe and senna.

Proteins

Some of the most potent toxins known are plant proteins. The seeds of the jumbie bean or rosary pea (*Abrus precatorius*) and the castor bean (*Ricinus communis*) have each caused many deaths. Other plant proteins such as the mitogens of pokeweed have proven to be very useful research tools.

Resins

Plant resins have many and varied effects on animal physiology. The resin from Cannabis is widely used (illegally) as a euphoric and narcotic, but it has also seen medical application in the treatment of glaucoma and the nausea associated with cancer chemotherapy. The urushiol from poison oak, ivy, sumac and the less-well known Christmas bush (*Comocladia*) is notorious for producing a vigorous contact dermatitis.

Gossypol from cottonseed oil and other plants has been shown to have potential as a male contraceptive by effectively decreasing sperm counts without lowering sex drive.

The terpenoids such as citrus oils are frequently favored as flavoring substances known in commerce as essential oils. Various other terpenes are skin irritants.

As we continue to investigate known plants and discover new species with unanticipated properties, the contributions of botanical biochemistry to modern society will increase.

PROPAGATION

Many of us are delighted to find interesting plants growing in their native habitat, but, understandably, many others want to bring these plants home to incorporate into a landscape plan. It is often impractical and frequently illegal to uproot plants from the seashore environment. While native plant nurseries are increasingly making many of these species of plants available, it is more fun to grow them yourself if time and resources permit. The science of plant propagation is the subject of many lengthy and detailed books which should be consulted by the serious grower. I have summarized here some of the techniques and considerations most relevant for the propagation of the plants to be grown in the environments discussed in this book.

Germination

Cleaned seeds of many species of plants may be stored for several years in sealed jars without loss of viability if kept in the cool and darkness of a refrigerator.

The seed coat or testa protects the embryonic plant but may stand in the way of artificial propagation efforts. The testa may contain or enclose germination inhibitors which must be dissolved and washed away before growth can proceed. The testa may also inhibit water absorption and gas exchange needed for growth. Ecologically the tough seed coat allows natural processes to disburse and bury the seed before the seed coat breaks down and allows germination.

The term scarification refers to various methods used to weaken the hard and impervious testa of seeds, enabling the embryonic plant to immediately absorb water, germinate, and begin growth. Successful techniques include acid baths, abrasion (sandpaper or file), heat (boiling water or torch), or cracking with a hammer.

The need for scarification may be avoided if seeds are harvested and planted when fully mature but before the testa toughens. The ease and rapidity of germination using this technique is often astonishing.

The seeds of some species of plants require light for germination, while others refuse to germinate unless they are completely buried in the soil. Others require several days of high temperatures or a strong daily fluctuation in temperature. When working with an unknown species it is best to try a variety of treatments until the particular requirements are discovered. The general rule which seems to work best for most plants is to bury the seeds as deep as their own width in a mixture of 1/3 carbonate sand, 1/3 good topsoil and 1/3 peat. Water daily or at least often enough to keep the soil moist. On islands where peat is an expensive import, fresh or composted grass clippings work just as well. Open shade is more desirable for seedlings than the full force of the fierce tropical sun.

Many salt-tolerant plants seem susceptible to damping-off and other fungi when grown in the nursery. It is my hypothesis that many of these pathogens cannot live in salty environments and thus there has been no selection for disease resistance in host plants. In addition to or instead of using fungicides with these plants I suggest the addition of a small amount of salt to the potting soil or the irrigation water. However, seed germination is generally sensitive to salt and frequently will fail to take place in very saline soils which may support the growth of more mature plants.

Discontinuous germination, in which individual seeds take extremely varied amounts of time to germinate, is common in many colonizing species of plants. Thus only a few of the seeds in the soil will germinate at any one time. This maximizes the probability that on a given site one or more individuals will find circumstances suitable to grow into a reproductively mature plant. When propagating one of these species it is wise to plant the seeds in a flat and transplant the individual seedlings to small pots as they mature. Extended patience will usually produce a reward of many more seedlings.

12

Cuttings

Cuttings of the vegetative parts of plants are often used by horticulturists to propagate new plants. An advantage of this form of propagation is that the new plants will be clones (genetic duplicates) of the donor plant, retaining all of its foliage and flower characteristics. The mixing and alteration of parental traits in progeny that occurs with sexual reproduction by seeds is never possible with cuttings, but chance desirable mutations such as the silver leaf in buttonwood can be retained indefinitely by vegetative propagation. For plants that produce few seeds or whose seeds have a poor germination percentage, cuttings are the only practical way to propagate the plant in the nursery. Cuttings often have the additional advantage that they bloom and fruit in a shorter time period than the equivalent age seedling.

Cuttings should usually be taken from the young but woody branches of the plant; however, extreme exceptions do occur. Young soft growth may respond best to rooting attempts. Almost any fragment of soft *Portulaca* stem can be induced to root. More mature hardwood cuttings may root readily. An example is the fence posts of gumbo limbo (*Bursera*), which often root with no additional care.

As soon as the cutting is taken, most of the leaves should be removed to reduce water demand. Cuttings should generally be kept moist by being wrapped in wet paper or cloth, then placed in plastic bags out of the sun until planted. Frangipani and cacti are two exceptions in which the cut ends should be allowed to dry thoroughly before planting. Many horticulturists routinely use various commercially available rooting hormones, but others eschew them except in special cases. Several different chemicals and mixture are available for promotion of rooting. The most widely used and generally successful is IBA (indolebutyric acid), which usually appears as a powder mixed with talc. The end of the cutting which was toward the root of the plant should be moistened and dipped into the rooting powder. An adequate amount of the rooting hormone will adhere to the stem which should immediately be set in soil. Although the wound produced by making the cutting is usually sufficient to induce rooting, a good thump with a hammer on the root end sometimes provides the trauma required to convince difficult plants to root.

Soil for rooting cuttings should always be light and loose of texture with a high sand content, to allow the easy penetration of tender rootlets, and some peat or compost to provide nutrients and water-holding capability. Although this soil will drain very well, it should not be allowed to dry, as this will cause new roots to shrivel. When root development is announced by new leaf growth, the cuttings should be placed gently into individual pots until they become vigorous enough to be planted out.

Air layering

Air layering is a technique which induces a plant to develop roots on one or more of its branches. At a distance of 1 to 2 feet from the end of a branch, a band of bark an inch or so wide should be removed from the stem. The white exposed wood should be dusted with rooting hormone and then covered with a handful of wet sphagnum moss, overlapping the bark on both ends. This soggy poulticed wound should then be wrapped with clear plastic sheeting such as plastic food wrap and sealed to the branch on each end with adhesive or electrical tape. Aluminum foil makes a more durable outer wrapping but does not allow the visual inspection of root development provided by clear plastic. After about 2 months the branch with its new roots may be cut from the parent tree and started in a pot to allow further root development before setting out. Keep in mind that this branch has been drawing water and nutrients from the parent plant up until the time of amputation and thus should be well watered and protected from strong sunlight until it shows new growth.

Irrigation

In addition to being able to develop an attractive seashore landscape, planting salt-tolerant species allows the use of brackish water for irrigation. When watering with brackish water, use enough water to insure that salt is leached away and does not accumulate near the soil surface as water evaporates. When using fresh water or treated sewage effluent for irrigation, remember that many seashore plants do poorly with excess water or fertilizer.

ESTABLISHING SHORELINE COMMUNITIES

Success in establishing new shoreline plantings depends on the careful planning and thoughtful execution of the following steps:

- Examine shorelines in your area with an environment similar to the one you intend to plant and record the species present.
- Draw the site you intend to plant and mark the ecological areas suitable for each species.
- Choose the specific species you would like to use. Remember a mixed-species planting is more likely to survive and adapt to a specific site.
- Distribute the plants randomly within their ecological zone. This yields a natural-looking community and disguises the inevitable unevenness in growth rate due to small variations in habitat conditions.
- Arrange plantings close enough to offer some mutual support against wind erosion, wind buffeting and foot traffic.
- Grow or buy a sufficient number of plants to firmly establish the community structure you are seeking.
- Dig a hole in proportion to the expected future root zone of the new plants and refill it with the same soil after it has been improved. On beaches with quartz sand, add some ground limestone and about 1/3 compost or aged manure. On clay soil, add about 1/4 carbonate sand and 1/3 compost or aged manure. On carbonate sand sites add half the volume of

compost or aged manure.

- Carefully examine your local annual climatic cycles and choose a planting time which is likely to offer more cool moist weather.
- If needed, use sand drift fences on dunes and mechanical support for woody species until the plants are established. The need for continued use of these aids is an indication of a mistake in plant selection for the habitat.

HOSTILE PLANTS

Many plants growing in dry environments have developed a chemical defense called allelopathy. This is the secretion of chemical inhibitors which have a direct or indirect effect on the growth of potential competitors. Some compounds influence only plants of another species while others hold fellow members of the same species in check. When you see evenly spaced plants in nature with bare ground around them you should suspect that they are secreting allelopathic polyacetylenes or terpenes which would make planting next to them a frustrating and unrewarding experience. In our area the common weedy tree *Leucaena leucocephala* is known to be so allelopathic that including its leaves in a potting mix will inhibit or prevent the growth of new seedlings of other species.

Sea Grape *Coccoloba uvifera (see page 44)*

Coast sandbur
Cenchrus incertus   Gramineae

Geographic Distribution   The genus of about 20 species is found in the warm parts of the world. This species and many similar ones occur in the southern U.S. and throughout the Caribbean basin.

Form   A bunch grass which is low growing with flat branching leaf blades. Seldom over 30 cm tall. All the sandburs are characterized by seed heads having stiff spiny burs.

Flower & Fruit Date   More prolific in the summer but in the warm parts of its range it will bear continuously.

Reproduction   The seeds are distributed when the burs adhere to passing animals. The clumps also extend vegetatively.

Propagation   Because of the production of sharp burs and the flat bunchy growth habit, this grass is generally considered undesirable and treated as a weed to be removed. Propagation is easily accomplished by seed or dividing clumps.

Habitat & Ecological Distribution Most often found on open sandy soil and many other places wherever the burs are dropped on disturbed soil. This plant shows high levels of asymbiotic nitrogen fixation which aids its growth on nutrient-poor soils.

Uses

Ornamental   The tufted seed heads and low growth habit are seldom appreciated due to adverse feelings of victims impaled by the burs.

Medicinal   Teas prepared from this grass have been variously prescribed as home remedies for kidney problems, fever, colds, vomiting and to induce lactation.

Edible An excellent forage grass before the burs form. The presence of burs in grass or hay renders the forage obnoxious to livestock.

Toxic The sharp spines on the burs have minute barblike bristles and can produce injury and infection when they puncture bare feet.

Ecological The plants are resistant to foot and vehicular traffic and are very hardy, thus they are useful for soil stabilization in heavy-use areas.

Notes   The generic name *Cenchrus* is derived from a Greek word for a type of millet. The species *incertus* translates as "unreliable." *Cenchrus tribuloides* is considered by some to be a synonym.

SALT MARSH CORDGRASS
Spartina patens   *Gramineae*

Geographic Distribution The genus has 16 species from Canada to Mexico, and in the West Indies and the Mediterranean. This species is found almost throughout the range of the genus.

Form Erect to spreading grass sparingly branched with very narrow (0.3 cm [1/8 in]) but long (45 cm [18 in]) leaves with margins rolled inward. The total height of the plant may reach a meter (3 ft). It forms continuous stands in coastal marshes but forms tufted bunches on coastal dunes.

Flower & Fruit Date The small inconspicuous flowers occur from June to October.

Reproduction The seed-head generally has 3 to 6 spikes and seeds establish new sites. Underground rhizomes spread *Spartina* vegetatively in established stands.

Propagation This grass seeds abundantly and often volunteers on suitable sites. Seed viability is low, usually in the range of 7 to 12%. Large-scale harvest of seed using standard agricultural methods is possible. The seed should be stored dry until it is planted on protected sites. Seeds may lose their viability when stored in a dry state for over a month. Light has been found to inhibit germination; thus, long submersion under muddy water may promote germination. On steeply sloping or exposed sites, salt marsh cordgrass can be transplanted to start new stands. Due to variation in strains, care must be taken to match the saltiness and water regime of the soil of the source plants and the site to receive the plantings. Plantings of vegetative material should be in the spring for best results. Stems rooted at the base, preferably with a section of rhizome attached, should be planted at depths of 12 cm (5 in) in 1-m (3-ft) rows with plants 1 m apart. Plantings in peat pots grow well and with their intact root system they survive better on difficult or salty sites. On dry sites, watering for the first few months aids establishment, but once established this grass is usually very persistent. On sandy sites, fertilizing with 8-8-8 or equivalent aids in the development of a robust stand. It

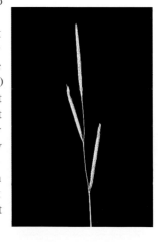

is probably most suited for moist salty sites.

Habitat & Ecological Distribution It is found on irregularly flooded salt and brackish marshes, saline shores, tidal creeks and sandy meadows from mean high water to the storm tide line. It often forms monospecific mats. Growth is best on muddy soils, with low competitive ability on clay or sandy substrates. Maximal salt tolerance for germination is 8% or about twice that of seawater. Under favorable conditions *S. patens* will grow from a few scattered clumps to a continuous meadow in less than 20 years. S*partina* plants have survived continuous submersion for 4 1/2 months and immersion for 23 1/2 hours per day in tidal cycles.

Uses

Physical Cattle are grazed on this grass and it is traditionally harvested as hay. Current agricultural research is developing cultivars which will produce increased yields of quality hay while being irrigated with seawater.

Toxic Under certain conditions it has caused prussic-acid poisoning in cows.

Ecological The detritus produced by this grass adds greatly to the productivity of the salt marsh. The colonies become so thick that they exclude other plants and greatly contribute to the prevention of coastal erosion.

Notes The genus is from the Greek meaning "a cord." The species name means "diverging" and is used to describe the growth habit.

Sea oats

Uniola paniculata   *Gramineae*

Geographic Distribution The genus of several species is found in the warm parts of the Americas. Sea oats are found from the southeastern U.S. to South America. Spike grass (*Uniola virgata*) is found in the Bahamas south through the Caribbean.

Form A tall coarse grass to a meter (3 ft) tall with seed heads to 2 m (6 ft). Usually found growing in clumps but may form a continuous cover given suitable time and habitat. The species *U. virgata* is similar in growth habit.

Flower & Fruit Date The attractive splayed seed heads of compressed spikelets

mature in late summer or early fall. The viability and number of seeds per spikelet is variable between pop-ulations. The seed head of *U. virgata* is an elongate, tightly compressed bundle.

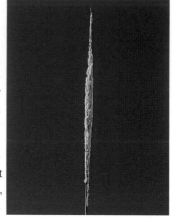

Reproduction Under natural circumstances reproduction is primarily by the growth and spread of rhizomes. The natural repro-duction by seeds is limited and irregular from year to year although under certain circumstances a dune will be liberally sprinkled with new seedlings.

Propagation Seeds should be planted shal-lowly in a mix of 3/4 sand and 1/4 peat with a bit of topsoil. If placed in full sun, germination will occur in less than a month. Although the higher organic content of the potting soil promotes growth, it may also lead to fungus problems. Thus a fungicide should be applied periodically in place of the weekly soluble fertilizer. Seedlings should not be set out until they are at least 30 cm (12 in) tall. The digging and dividing of a part of adult plant clumps is the more successful method of establishing sea oats. The plants should be dug as deeply as possible in order to include a portion of the rhizome. When planting, the stock should be set at least 30 cm (1 ft) deep and tightly packed in. The best planting time is wet weather, or the dormant period December to March. A balanced time-release fertilizer and modest watering can produce a lateral spread of clumps of up to 1.2 m (4 ft) per year. Soil-stabilizing fabric or snow fence aids establishment in rapidly moving sand.

Habitat & Ecological Distribution Sea oats flourish in areas where sand is drifting. When growing as isolated clumps they may trap large amounts of wind-blown sand, resulting in sea oat hummocks. They can grow up through as much as 60 cm (2 ft) of new sand per year. While clumps are initially slow to becomeestablished, they are very persistent under

circumstances of heat, drought, salt, and lack of nutrients. While admi-rably suited to the dry dune envi-ronment, sea oats become stunted and do poorly if the dunes are low enough in elevation to allow the water table to reach within 0.3 m (1 ft) of the surface. Prolonged inun-dation of the roots kills sea oats. Their salt tolerance is greater for salt spray than salt in the sand.

Uses

Ornamental The seed heads are much prized as components of dried flower arrangements.

Edible The seeds of both the above species may be cooked and eaten as a cereal.

Ecological Within their range sea oats are the most effective plant for stabilizing primary sand dunes against wind erosion. Although wave erosion may undercut the edge of established stands, they usually withstand the pounding and overwash of severe storms.

Notes *Uniola* is an ancient plant name. The species *paniculata* refers to the loose cluster of the head. The species *virgata* refers to the rodlike seed head. Maximum growth is obtained when daytime temperatures regularly exceed 80° F. Seeds from more northerly populations require a period of cold before germination will take place.

SEASHORE RUSH GRASS
Sporobolus virginicus   Gramineae

Geographic Distribution The genus of 150 species is found throughout the warmer parts of the world. This species was found originally from Virginia to Brazil. It is one of the most common seashore grasses in the West Indies and is now found worldwide.

Form A prickly grass with seed heads to 45 cm (18 in) tall. The leaves are in two vertical rows with overlapping sheaths.

Flower & Fruit Date Seemingly flowering at random in the warm part of its range.

Reproduction A narrow seed head 2 to 10 cm (1 to 4 in) long produces the many small seeds which drop free at maturity.

Propagation May be started by transplanting rooted rhizomes, dividing adult clumps, or by planting seed either directly in the ground or in small pots. Seed viability is high, usually in the range of 75%. Nitrogenous fertilizer accelerates the growth rate. Regular mowing promotes a thick stand and controls unwanted weeds.

Habitat & Ecological Distribution A coarse form of this grass grows on beaches beyond the reach of waves and on disturbed salty dry sites. A more finely textured form grows on salt flats, marshes, near mangroves, and on muddy shorelines. Both forms can grow in waterlogged

21

anaerobic soil. The young plants are adversely influenced by inundation and high salinity which mature plants tolerate with impunity. The species usually does best on moist silty sand.

Uses

Ornamental This grass is the candidate of choice for those needing a lawn on a salty site.

Edible This grass is often the only one which will grow on salty sites and can be irrigated by seawater. It is one of the few salt-tolerant grasses which provide pasture for cattle and deer. Total forage production is decreased with heavy grazing pressure. Cultivars for the production of hay are being developed.

Ecological Once established, this grass is very resistant to erosion as evidenced by its survival after pounding by hurricane waves. The seeds are consumed by birds and mammals. Deer prefer to graze on the fine form of this grass. *Sporobolus* has been found to be an integral habitat requirement for seaside sparrows. The hardy nature and fecundity of this grass even in the presence of elevated levels of arsenic, manganese, iron, and copper have made it useful in vegetating toxic mine tailings.

Notes *Sporobolus* is derived from the Greek *sporo,* meaning "seed" and *bolus,* meaning "to throw," the combination referring to the loosely attached seeds which are easily thrown from the head. The species *virginicus* refers to the plant originally being described from Virginia. As soil salinity increases above 10 parts per thousand, or if anaerobic soil conditions prevail, the growth rate declines and the grass becomes stunted.

SEASHORE SALTGRASS
Distichlis spicata   *Gramineae*

Geographic Distribution The genus is composed of 3 or 4 species restricted to the Americas. This species is found on both coasts of the U.S. south to Mexico and the northern West Indies.

Form A grass with a thin, tough horizontal rhizome, vertical stems to 18 inches and slim alternate trough-shaped leaflets in a single plane. It forms thick mats from underground rhizomes.

Flower & Fruit Date The flowers appear in late summer in the northern part of the range and in every month in the Caribbean.

Reproduction The erect spikelets are composed of leathery husks with 5 to 15 seeds per head. Vegetative reproduction is vigorous, with the young leaves arising from rooted rhizomes in a ropelike pattern.

Propagation Saltgrass seeds germinate best with a moist seedbed and diurnal temperature shifts of at least 35 degrees. The best germination for many populations is obtained when seedbed temperatures exceed 100 degrees on a daily basis, but highs of 87 degrees will usually produce modest  levels of germination. Division of clumps into plugs is the most successful method of establishing new cover. Bare root planting of rooted rhizomes into moist soil will also establish a stand. Other researchers have had the best success with stock rooted in peat pellets. Nitrogenous fertilizer after planting brings a rapid response. Saltgrass readily invades suitable sites; thus artificial introductions are seldom needed. Mowing stimulates a thick growth.

Habitat & Ecological Distribution This hardy grass is found on the edges of salt flats, marshes, and protected beaches. Frequently it is mixed with other species except on the poorly drained salty patches where it dominates. It can withstand regular seawater inundation and will continue to grow upward when exposed to modest sedimentation.

Uses

Ornamental This is one of the few grasses capable of being made into a lawn when exposed to seawater and heavy salt spray.

Edible When grown on salty soils as a cereal grain, the seeds are not salty and livestock graze on the foliage. The grain has an abundance of all the amino acids required for human nutrition except for methionine and cystine. It contains more bran and fiber than wheat or barley and about the same amount of starch. The flour is suitable for breadmaking. A cultivar to produce hay for livestock feed while being irrigated with salt water has been tested and is being further refined.

Physical This grass can be used to stabilize mine tailings and bauxite residue, paticularly when the substrate is supplemented with sewage sludge.

Ecological Established stands form a very effective barrier to erosion. Birds and mammals eat the seeds. Members of this genus are being promoted for food production on salinized agricultural soils.

Notes *Distichlis* is from the Greek word meaning "two rows." The species *spicata* refers to the sharp tips on the leaves. In laboratory experiments this is one of the few plants which grows better in a salty environment. Maximum growth rate has been obtained in soils with 15 parts per thousand salt content, which is about half the salinity of seawater.

Turtle Grass

Thalassia testudinum   *Gramineae*

Geographic Distribution The genus is found only underwater in the Indian, Pacific and Atlantic Oceans. This species is widespread only in the tropical western Atlantic.

Form Leaves are ribbonlike, 6 to 12 mm (1/4 to 1/2 in) wide and up to 30 cm (1 ft) long, arising from fibrous rhizomes. Roots may penetrate up to 4 m (13 ft) of sediment.

Size Forms varying-sized patches of grass meadow with individual blades typically less than 15 cm (6 in) above the sediment.

Flower & Fruit Date Flowers and fruits in warming water temperatures after the winter minimum.

Reproduction Turtle grass reproduces both vegetatively by growth of rhizomes and sexually by seeds distributed by water currents.

Photo: Vance Vincente

Propagation Transplantation by plugs of sediment with intact plants is usually most successful. Bare rhizomes anchored in suitable areas have also established successful colonies. The use of iron in anchoring results in disintegration of the plants. Fruits may be collected by scuba divers and dehisced by immersion in fresh water. Planted in small pots in quiet water, the seedlings do well when transplanted.

Habitat & Ecological Distribution Turtle grass is the most abundant of the "seagrasses" in the warm, clear, shallow waters of the western Atlantic. Natural beds of *Thalassia* have suffered greatly from man's activities. Dredging and anchoring have directly eliminated the self-supporting mats of rhizomes which maintain the integrity of the community. Sediment loads caused by poor land-management practices, pollution and turbidity induced by boat traffic have all reduced the light penetration needed for vigorous plant growth.

Uses

Edible Sheep have been found to have increased weight gain when their diet is supplemented with up to 10% turtle grass.

Physical Farmers make use of windrows of turtle grass collected from beaches after storms as mulch and fertilizer for gardens.

Ecological Turtle grass is used as food by green sea turtles, manatees, herbivorous "reef" fish and conchs. Many species of invertebrates consume the dead leaves and other detritus in grass beds. Established beds of turtle grass stabilize near-shore sediments. The interstices between the blades and roots of turtle grass provide nursery areas for shrimp, lobster, conch and many important fish species.

Notes *Thalassia* refers to the plants' growth in the sea. The species *testudinum* refers to the turtles which depend on the grass as a food source. Natural regeneration of seagrass beds disturbed by boat anchors and storms is slow. When anchoring a boat, it is a good conservation practice to place the anchor in sand patches rather than grass beds. The second most common seagrass is *Syringodium filiforme* (manatee grass) with wirelike leaves often growing interspersed with *Thalassia*.

COASTAL SEDGE
Cyperus planifolius   *Cyperaceae*

Geographic Distribution The genus has 900 species worldwide. This species is found in South Florida and the West Indies.

Form A grasslike herb growing in clumps to 0.6 m (2 ft) tall with seed heads on triangular stalks to 1 m (3 ft).

Flower & Fruit Date The tiny inconspicuous flowers occur in every month.

Reproduction The elevated seed heads allow distribution of the seeds by wind threshing.

Propagation Clumps of sedge may be subdivided for planting. The seeds require warm temperatures and light for germination.

Habitat & Ecological Distribution Almost always near salt water on sandy beaches, in damp spots behind dunes, on clay pans, and on rocky shorelines.

Uses

Ornamental The seed heads have been used in dry flower arrangements.

Ecological This sedge forms the dominant cover on many small islands and thus provides shade and concealment for many species of nesting seabirds.

Notes *Cyperus* is an ancient Greek name for this group of plants. The species *planifolius* refers to the flat leaves. It is also called sand or coast sedge.

Spike rush

Fimbristylis cymosa ssp spathacea   *Cyperaceae*

Geographic Distribution The 200 species in the genus are found world-wide. This species is found in peninsular Florida, Mexico and the Caribbean and also in the Old World tropics.

Form A grasslike herbaceous plant with leaves arising from a central clump. The leaves reach about 30 cm (12 in) high with the central stalk (culm) extending another 15 cm (6 in).

Flower & Fruit Date The inconspicuous flowers form at the tip of the culm in all months.

Reproduction The spikelets on the seed head are oval and less than 6 mm (1/4 in) long.

Propagation Entire clumps may be transplanted or subdivided.

Habitat & Ecological Distribution Found growing on brackish soils bordering mangroves, salt flats, tidal creeks and moist to dry dunes.

Uses

Ornamental This sedge has been used as an accent or border plant on moist salty sites.

Medicinal Related species have been used in folk medicine to treat the umbilical wound, and as a remedy for seasickness.

Physical Historically, paper has been made from Fimbristylis stems.

Notes *Fimbristylis* is from the Latin words meaning "fringed style." The species *cymose* is derived from the frequently cymose inflorescence

Coconut palm

Cocos nucifera   *Palmae*

Geographic Distribution The genus has only one species whose geographic origin is debated. The coconut was unknown on the shores of the Atlantic at the time of the first European explorations. Today it is found worldwide on tropical shores and it is widely planted inland at low elevations.

Form The graceful palm typical of illustrations of idyllic tropical beaches. To

20 m (65 ft) tall and 50 cm (20 in) in diameter at the thickened base.

Flower & Fruit Date The small light-yellow flowers of each sex are borne year-round on stalks up to 1.3 m (4 ft) long.

Reproduction The large nut is a single seed which takes about a year to mature on the tree.

Propagation The coconut is best propagated by laying the nut on its side in a shallow hole leaving about 1/3 of the nut exposed above the soil. The palm should be transplanted at the age of 7 to 9 months. Only dwarf and other strains resistant to lethal yellowing disease should be planted. The rate of growth is greatly enhanced by the amount of water and nutrients available. The coconut requires 6 to 10 years before it bears fruit and may not reach full production for 20 years. Healthy trees will produce 50 to 100 coconuts per year and live 80 to 100 years. There are many different cultivars with varying growth habits, environmental preferences and different size, shape and structure of the nuts.

Habitat & Ecological Distribution The coconut prefers a light, sandy, well-drained soil and is quite salt-tolerant. It cannot tolerate freezing temperatures. The trees do well in low areas where the water table is relatively high provided there is sufficient fluctuation to allow aeration for root development.The fibrous roots extend laterally at least as far as the fronds. Once started, the trees can survive in an astonishing variety of adverse conditions from heavy laterite soils to almost pure sand with extreme heat and wide variations of pH.

Uses The coconut is possibly the most broadly useful tree in the world.

Ornamental It has been widely planted for its modest shade and very attractive appearance.

Medicinal Coconut oil is used as a vehicle for many skin medications. Water from the ripe nut is a diuretic and is sufficiently sterile that it has been used as a fluid for direct injection into the veins of medical patients. The meat from the unripe nut is a laxative. Over 60 folk medicinal uses have been prescribed for various parts of the tree treating a broad spectrum of ailments. Modern medicine has found that the short molecular chains of the oil make it easily absorbed by individuals with fat-absorption problems. The low viscosity fractionated oil is a useful medium for oral administration of fat-soluble medications.

Edible The liquid from the center of both ripe and unripe nuts is used as a beverage. The flesh of the nut is eaten fresh (a firm jellylike consistency from immature nuts or the coconut of temperate commerce from mature nuts) or dried and sold as copra. A germinated nut becomes filled with a spongy substance called bread which may be eaten fresh or toasted over a fire. Oil extracted from the nutmeat is used in cooking, margarine, ice cream and many other foodstuffs prepared both industrially and in the home kitchen. The cake remaining after oil extraction from copra is a rich livestock fodder. The sap from severed flower stalks can be consumed fresh or fermented into an alcoholic beverage called toddy. Toddy can be allowed to become vinegar by further fermentation or it can be distilled into the potent alcoholic beverage arrack.

Physical On resource-poor tropical islands coconut leaves provide thatch for roofs and are woven into wall lattice, fish traps and baskets. The trunk is used as construction timber and for furniture manufacture. The fibers in the outer husk of the nut are called coir. It is used as padding (in mattresses, furniture and life jackets), mats, ropes, brushes and filters. The shells are used as utensils and make an excellent fuel source. Oil from the meat is used in shampoo, soap, hydraulic fluid, paint, synthetic rubber, plastics and many cosmetics. The roots provide a dye.

Notes The genus is named for a Portuguese word for monkey which refers to the face on the end of the nut. The species is from the Latin word meaning "nut bearing." Because the nut can stand long seawater immersion, it has drifted and become established on many remote shorelines. The abundant tough roots keep the tree standing in the most fierce hurricanes. Many people on small islands have survived catastrophic storms by lashing themselves to coconut trees. Bees produce a light mild honey from the nectar. One hundred g (4 oz) of coconut meat contains the following: 346 calories, 3.5 g of protein, 35 g of fat, 9.4 g of carbohydrate, 4 g of fiber, 13 mg calcium, 95 mg phosphorus, 1.7 mg iron, 23 mg sodium, 256 mg potassium. The same amount of the liquid from the interior of the coconut contains: 22 calories, 0.3 g protein, 0.2 g fat, 4.7 g carbohydrate, 20 mg calcium, 13 mg phosphorus, 25 mg sodium, 147 mg potassium. Coconut oil extracted from the meat is composed of triglycerides of lauric and myristic acids with lesser quantities of caprylic, caproic, oleic, stearic, and palmitic acids.

SILVER PALM

Coccothrinax argentea   *Palmae*

Geographic Distribution The genus has 30 species distributed in the Caribbean area. The silver palm occurs in South Florida and Cuba. The very similar tyre palm (*C. alta*) is found naturally in Puerto Rico and the Virgin Islands and has been introduced in Florida.

Form A very slender palm up to 12 m (40 ft) tall with a smooth trunk to 12 cm (5 in) showing leaf scars as faint rings. The leaves are shiny green on top and silvery white on the bottom, flashing as they are rotated by the wind.

Flower & Fruit Date The sprays of fragrant white flowers occur in late spring. The clusters of dark purple 9 mm (3/8 in) fruit mature in the fall.

Propagation The fleshy outer cover should be removed from the seed. It should be dried several days, then dusted with a fungicide. Plant in a well-drained but moist medium (perlite is good), then cover with 1/4 inch of shredded sphagnum. Soil temperature should remain above 80° F. Germination will take 50 to 75 days and in extreme cases as long as 8 months. An alternative method is  to soak seeds in warm (100°-120° F) water for 3 weeks, then plant in moist sand maintained at above 80° F. Seeds older than 3 months have a greatly reduced germination rate. Fertilizing with a balanced fertilizer accelerates growth and produces a better looking plant. Growth rate is slow under all circumstances. It may take 2 years for a seedling to be large enough to require transplanting from a 2-inch pot and another 50 years to produce a 3-m (9-ft) tree. With adequate water and fertilizer on good sites in full sun, 3-m trees have been produced in 5 years. Transplanting trees from the wild is difficult and seldom meets with success.

Habitat & Ecological Distribution This palm is salt-tolerant and can be found growing in inhospitable rock clefts with minimal soil. It is sensitive to frost.

Uses

Ornamental Because of its small size and attractive foliage, it is used as an ornamental, particularly on poor sites. Nursery or home-grown plants should be used for landscaping. Do not dig wild trees as they will probably die and it will take natural regeneration many years to replace them.

Edible The fruit is edible and has been made into wine. The terminal bud or cabbage can be used in a "hearts of palm" salad, but should be harvested from the wild only in emergencies.

Physical The Caribe Indians used the wood to make bows. The leaves were used as thatch and the petiole can be split and woven while moist into ropes, baskets and hats. When bundled together the leaves make durable brooms.

Ecological White-crowned pigeons and other fruit-eating birds find the berries very attractive as a food source. They also serve as the primary agents of seed dispersal.

Notes The genus is from the Greek meaning "berry-fan" in reference to the spreading fruit clusters. The species *argentata* refers to the silver undersides of the leaves. The species *alta* refers to the tall thin growth habit. Some authors consider the two palms to be variations of the same species. The taxonomy of this genus is such that these Caribbean palms have been known by several names. Other common names include bay top, thatch palm and teyer palm. The latter is probably derived from the practice of using strips of the leaves to tie bundles and lash small items together.

ALOE

Aloe vera   *Liliaceae*

Geographic Distribution The genus of over 200 species occurs predominantly in Africa. This species, originally a native of the Mediterranean coast, is now widely introduced and naturalized in the tropics and subtropics of the world.

Form A perennial herbaceous plant with a short and inconspicuous stem. The plant is composed of a cluster of thick, turgid, tapered, toothed leaves to 0.5 m (20 inches) in length. The leaf is structured in 3 layers: the outer layer is a leathery cuticle, which overlays a vascular area containing most of the starch and chlorophyll. The central bulk of the leaf is an area of mucilage-containing cells.

Flower & Fruit Date The yellow to orange bell-shaped flowers are borne densely in the winter or spring on the upper parts of a stalk up to 1 m (3 ft) tall.

Reproduction The numerous black seeds are born in a leathery capsule which splits open when ripe.

Propagation The seed will germinate after about 4 weeks of planting in a light sandy soil. Mature plants send out suckers that are easily transplanted. Aloe does best in open shade but easily tolerates full sun. While it will persist on poor dry soil, its slow growth rate responds well to water and fertilizer.

Habitat & Ecological Distribution Frequently escaping and persisting after cultivation, aloe is very salt- and drought-tolerant and may be found on rocky shorelines, dunes or intermingled with other vegetation on almost every soil type.

Uses

Ornamental It is widely planted as an ornamental, both for its appearance as a hardy succulent and for the attractive flower stalks. Numerous other showy species of aloe are becoming increasingly available from nurseries. Most do very well in rock gardens and as ground covers on salty sloping sad sites.

Medicinal Folk medicine has made use of this plant to treat an almost unending list of external and internal afflictions since the 4th century B.C. It has also been used as an aphrodisiac tonic and to promote the development of the mammary glands. When the juice which drains from the outer vascular areas of the cut leaves of the plants is concentrated and solidified it becomes the commercial medical purgative product "aloin" or "aloe-emodin" previously officially recognized in both the British and American pharmacopoeias. The cathartic effect is produced by a group of chemicals called anthraquinone glycosides of which barbaloin makes up 70% of the total. Its bitter taste has brought it into use to promote weaning by being rubbed on the breasts and on the fingers to break a sucking habit. The fresh mucilaginous pulp and gel extracted from the central part of the plant is the commercial "aloe vera." This product has been demonstrated in clinical tests to promote and accelerate healing in thermal tissue injuries such as frostbite and burns. The mode of action is to inhibit the formation of tissue-injuring thromboxanes at the site of the injury. The presence of chrysophanic acid has been used to explain the dramatic effectiveness of aloe gel in treating abrasive injuries to the skin and its general promotion of healing. Various extracts have shown anticancer activity.

Edible The leaves are edible. The pulp is used as a flavoring ingredient in many different foodstuffs including alcoholic and nonalcoholic beverages, frozen deserts, baked goods and confections.

Toxic The normal dosage is about 2-5 grains of dried emodin but doses over 10 grains are likely to induce abortion due to pelvic congestion.

Physical The gel squeezed from the leaves is used in skin-softening and moisturizing lotions, suntan lotions, cosmetics, shaving cream and shampoo.

Ecological The tubular flowers are much favored by hummingbirds.

Notes This plant has also been called *Aloe barbadensis*. The Latin *aloe* is derived from the Arabian name *alloch* or *alloeh*. The Latin *vera* means "true," and the *barbadensis* refers to the island of Barbados which has been a commercial source of the product since the 17th century. The cathartic emodin is the chemical trioxy-methyl-anthraquinone $(C_{15}H_{10}O_5)$. The juice inhibits the bacteria *Salmonella paratyphi, Escherichia coli, Serratia marescens, Enterobacter cloacae, Klebsiella pneumoniae, Pseudomonas aeruginosa, Corynebacterium xerose, Staphlococcus aureus, Streptococcus pyrogenes, S. agalactiae, S. faecalis, Bacillus subtilus, Candida albicans,* and *Citrobacter species*. The thick stringy mucilage-like texture of fresh aloe gel is due to the presence of polysaccharide glucomannans and pectins. The gel also contains mannose, glucose, 17 amino acids, glutamic, malic, succinic, and citric acids, 4 sterols, a sapogenin, quinones and anthraquinones, the enzymes cellulase, carboxypeptidase, bradykininase, catalase, amylase, and an oxidase. With the above and many other unlisted compounds present in the gel it is no wonder that many of the beneficial effects are as yet unexplained.

SPANISH BAYONET
Yucca aloifolia   *Liliaceae*

Geographic Distribution There are 30 species of yucca in the warm and arid areas of the Americas. This species is found in Florida, Mexico and the Caribbean.

Form An upright plant composed of broadly branched thick stems to 6 m (20 ft) high with rosettes of stiff, sharp-pointed, daggerlike leaves at stem tips. The bottom leaves die and form a beard as growth proceeds, eventually dropping off to expose the trunk.

Flower & Fruit Date The creamy white bell-shaped flowers, fragrant at night, usually form dense sprays on stalks in the summer.

Reproduction The blackish-purple fruits with many flat black seeds mature in the fall.

Propagation The best method for starting yucca is transplanting of young suckers from the base of mature plants. If material is available the trees may be started by 3-foot sections of tops cut off with the butt ends planted about a foot deep. The base of living leaves should not be placed below ground level. After

planting, the bottom 2/3 of the green leaves should be cut off. The yucca will also grow well from seed pressed into the soil surface and covered with a light sprinkle of sand. In many areas artificial pollination of the flowers is required to produce seeds. The leaf spot that sometimes occurs on this plant is usually caused by *Coniothyrium* and can be controlled by commercial fungicides.

Habitat & Ecological Distribution The yucca thrives in dry sandy soil and can survive in many rocky dry habitats. It is very salt- and drought-tolerant as well as being wind resistant.

Uses

Ornamental The distinctive appearance has frequently been utilized to provide a hardy, low-maintenance accent or border plant on dry sites. Dense clumps can be used to protect more tender plants from salt spray and wind buffeting. Because of the fierce spines on the leaf tips they should not be planted close to paths or play areas. It thrives as a pot plant and continues to look good with severe neglect. A bit of water and, more rarely, fertilizer accelerates growth rate.

Medicinal This plant is a source of the steroidal *sapogenin Sarsapogenin*. Extracts of the rhizomes have shown both anti-inflammatory and oxytocic activity.

Toxic extracts have proven to be toxic and of potential use in the control of the mollusc *Lymnaea cailliaudi,* which is an intermediate host of a human liver fluke.

Edible The flowers are edible raw in a salad and are sometimes fried in a batter. The fleshy fruit is edible but bitter. The flower stalk may be peeled, cooked and eaten. An extract from yucca stem is used as the foaming agent in root beer and other nonalcoholic frothy drinks.

Physical The fibrous leaves have been used in brooms, brushes, baskets, belts, cordage and weaving. In World War I, 80 million pounds of fiber were used to make burlap for bags and similar uses. The low-growing species makes an inconspicuous but effective barrier to foot traffic.

Ecological Yucca will both stabilize shifting sand and serve as a windbreak if closely planted in elongate clumps at right angles to the wind.

Notes The Haitian Indian name for the plant was used by Linnaeus when he named the genus *Yucca,* which is often used as the common name. The species *aloifolia* refers to similarities of the leaves to those of the aloe plant. The yucca is pollinated by a small white symbiotic moth (*Pronuba yuccasella*) whose larvae feed on yucca seeds. Two *spirostanol saponins* have been isolated from the rhizomes. Several steroid glycosides designated as yuccaloside A, B and C have been extracted from the leaves. The leaves are very rich in tigogenin and may be harvested, dried and stored for over 4 years without significant change in concentration.

CENTURY PLANT
Agave missionum   *Agavaceae*

- **Geographic Distribution** The genus has 300 species distributed in the dry areas of the tropical Americas. Many of the species have been cultivated and moved about. This one is found in Puerto Rico and the Virgin Islands.
- **Form** The plant grows as a rosette with sharply pointed, sword-shaped, thick leathery leaves with a hard brown spine on the tip. As with many drought-adapted plants, the roots form a dense mat just below the soil surface to allow maximum absorption of even the slightest rain. Prior to flowering

the plant may reach a height of 6 feet.

Flower & Fruit Date The flower stalk growing at the rate of 5 cm (2 in) per day emerges in March or April. The 30 to 40 horizontal branches flower progressively up the stalk, producing a magnificent golden cone.

Reproduction The plant takes many years to accumulate the reserves necessary to produce the 6 m (20 ft) flowering stalk. After flowering once, the plant dies. A few pods with seeds form on the stalk but many small plants (bulbils) may form after the flowers are dropped. These eventually drop to the ground and readily take root. Some plants reproduce vegetatively by sending out rhizomes which give rise to suckers.

Propagation Suckers from parent plants can be moved, but the bulbils which sprout and begin growing while on the flower stalk are most easily propagated in pots or directly in the soil on a dry site.

Habitat & Ecological Distribution The agave is able to survive on steep rocky sites with minimal soil and rainfall.

Uses

Ornamental Makes an attractive and durable border plant requiring minimal care in dry areas. Agaves make harmonious companions to cacti and other succulents.

Medicinal The juice from the leaves has been used in home remedies as a laxative, diuretic, and to treat dysentery, jaundice, sores, sprains, leprosy, syphilis and skin problems. The plant is a source of steroidal

sapogenins. One of these, hecogenin, commercially extracted from *A. sisilana*, is a source of cortisone and other pharmacologically active compounds. Diosgenin from agave was the original source material for the manufacture of oral contraceptives.

Toxic Some people develop a severe rash after contacting the juice, and others exhibit a slow-to-heal injury when impaled by the terminal spine. Sensitive individuals may show an allergic reaction to the fiber used in mattresses.

Edible The drink aguamiel is the fresh juice of the agave flower stalk. When fermented it becomes pulque and when distilled it iscalled mescal. The flowers are edible, and the central bud of many agaves is cooked and eaten as a vegetable.

Physical The long, strong, coarse fibers in the leaves have been used to make brushes, baskets, nets and ropes. The commercial fiber sisal is produced from *A. sisilana*, which is also widely planted as an ornamental. Henequen fiber is produced from *A. fourcroydes*. The pulp waste from henequen fiber production is fermented into alcohol in Mexico. The juice from the leaves impregnated into wallpaper renders it immune to termites. The juice also has detergent properties. The cuticle contains a hard wax similar to carnauba.

Ecological As an extremely drought-resistant species it helps stabilize the soil on dry slopes. Birds, bats and many insects feed on the nectar of the flowers.

Notes The name Agave is used both as the genus name and the common name for this group of plants. It is derived from an ancient word meaning "noble." The common name century plant refers to the common myth that it takes a century for the plant to mature and bloom. That may in fact be true of plants growing in tubs in nothern climates. When exposed to the tropical sun on a good site, the life cycle may be completed in 20 years. The juice contains pectin, vitamin C, and up to 11% of sugars which could be fermented. The steroidal sapogenins tigogenin, gitogenin, neotigogenin, sarsapogenin, sisalogenin, gloriogenin, gentrogenin, and yamogenin vary in concentration in the leaves, flowers, flower stalk, bulbils and rhizomes. As the plants grow older the sapogenins increase in concentration in the juice but contain successively fewer hydroxyl groups.

SPANISH MOSS

Tillandsia usneoides   *Bromeliaceae*

Geographic Distribution The genus of about 500 species is found in warm areas of the New World. Spanish moss is found from Florida to Chile.

Form The epiphyitic *Tillandsia* forms masses of silver-gray vegetation on almost anything offering support in the sun. The color is produced by a covering of minute velvety scales covering the stringlike leaves. The plant has no roots as an adult but absorbs water over its entire surface through the scales. Spanish moss branches alternately right and left at the nodes, and after reaching a length of about 20 cm it begins to die at the base as it grows at the tip. New plants are produced when the central stalk connecting side branches to the terminal shoot die. The long beardlike streamers, typically found on oaks and cypress, are formed by overlapping colonies of shorter plants. Ball moss (*T. recurvata*) forms compact clusters of a similar appearance.

Flower & Fruit Date Each plant produces one inconspicuous flower per year which releases its fragrance at night in the warm months.

Reproduction The seed capsule opens in the winter and some of the 11 to 20 tiny seeds germinate in place. The seeds and seedlings have a mass of fine barbed hairs which serve as a parachute for wind dispersal and aid in the attachment to a suitable substrate. Initially the new seedling produces adventitious roots to help hold it in place, but these are not present in the mature plant.

Propagation Spanish moss can be grown from seed, but simple vegetative propagation by tying a clump in place until it takes hold is most easily accomplished. While Spanish moss can survive indoors for months in moderate humidity without water, rainfall or regular watering along with some sunlight are needed for healthy vigorous growth. Rare sprinkling with dilute liquid fertilizer also aids growth.

Habitat & Ecological Distribution As a plant not requiring any soil, *Tillandsia* is found in a great variety of habitats which support it in an elevated position. It is tolerant of a wide range of temperatures and will grow in full sun or dense shade by making physiological changes without a change in form. High relative humidity at night is needed for growth.

The mineral content of Spanish moss and ball moss growing on trees or nearby electric wires is not significantly different. Its growth rate, distribution and abundance are enhanced by the availability of leachable nutrients in host trees and the amount of runoff which flows on the trunk and branches rather than dripping from the leaves. Cypress and oak have high levels of Ca, Mg, K, and P supporting dense draperies of moss, while pines have much lower levels of nutrients and sparse moss. Allelopathic factors in the bark of host trees may also influence distribution. Spanish moss seems to do better on dead trees due to the increased availability of sunlight and nutrients released by decay. The mineral and bark characteristics of the host tree are more important to the establishment and growth of moss than light and water. While it is common on the coast it is not tolerant of much salt spray. Its very spotty distribution on both large and small scales seems to indicate a limitation by dispersal factors rather than by biologically suitable growth conditions.

Uses

Ornamental Long beards of Spanish moss draped from the broad spreading branches of mature trees give them a venerable and stately appearance.

Medicinal As a home remedy, a tea prepared from Spanish moss has been used to treat female and gallbladder complaints. A tea prepared from the moss has long been used as a folk remedy for diabetes in Louisiana and experimentally has been found to reduce blood glucose in rats 4 to 8 hours after ingestion. Future research will be needed to investigate its potential as a drug source for treating diabetics. Spanish moss was believed to promote healing and was used as a bandaging material by American Indians and by soldiers in the American Civil War. The presence of the proteolytic enzyme bromelian is thought to break down dead tissue and fibrin deposits to enhance the rate of healing. An extract of Spanish moss has been shown to have significant analgesic effects in laboratory experiments. The Mayo Clinic has found that sterilized Spanish moss used in surgical dressings will absorb more water than the equivalent weight of cotton. Other studies have shown it to have antibacterial and estrogenic effects.

Toxic Spanish moss used as furniture and mattress padding and repeatedly exposed to moisture may support the growth of *Aspergillus* or *Fusarium* which in turn produce aflatoxins and trichothenes. Any manufactured product using Spanish moss should be kept out of the rain to avoid accidental production of these liver toxins.

Edible Spanish moss has been used as a component of livestock fodder and has been found to contain significant amounts of carotene and ascorbic acid.

Physical Spanish moss has been used as a springy filler in upholstery and mattresses. Three compounds capable of showing the liquid crystal phenomenon have been isolated and identified from Spanish moss.

Ecological Many birds, rodents and one species of bat use thick growths of the moss as a nest site or use bits of moss in their nests. Over 160 species of insects plus many mites and spiders have Spanish moss as their primary habitat. It seems that Spanish moss may be a biological indicator of atmospheric lead.

Notes The genus is named for the Swedish botanist Tillands. The species *usneoides* is named for the lichen genus *Usnea,* which it resembles. The species *recurvata* is from the Latin word meaning "curved backward." The chemistry of the foliage is very complex with many free and esterified triterpenes, sterols and flavone glycosides. Palmitic, stearic, lauric and myristic acids predominate as components.

PINGUIN

Bromelia pinguin   *Bromeliaceae*

Geographic Distribution The genus with many showy species is found in the American tropics. This species occurs in the wild from Mexico to Panama and the Caribbean islands. It is cultivated in Florida and the Bahamas.

Form A herbaceous perennial composed of a rosette of long, narrow, leathery leaves to 1 m (3 ft) tall and 1.5 m (5 ft) wide, resembling the related pineapple. The dark green leaves arise one at a time in a spiral from the center of the plant with about 120 degrees between leaves. The cross section of each leaf forms a trough

which channels water and organic nutrients to the center of the plant for absorption. The leaves are bordered by hooked, stout, sharp savage spines. The youngest inner leaves turn reddish when a flower stalk is developing.

Flower & Fruit Date The inconspicuous little pink flowers open sequentially a few at a time among the whitish bracts of the tall central spike.

Reproduction Vegetative reproduction is most common with the spreading

underground rhizomes starting new plants 25 to 75 cm (10 to 30 in) from the parent. The fuzzy yellow globular fruits, 2 to 3 cm (1 in) in diameter, are borne on the central stalk. Each spike has 10 to 75 fruits and each fruit has 30 to 50 shiny black seeds embedded in the sweet juicy pulp. The plant is well suited for having its seeds dispersed after consumption by large mammals. After fruiting, the plant ceases to add new leaves and gradually dies while it continues to produce offshoots. When the red land crab *Gecarcinus lateralis* is present in abundance it has been shown to eat virtually all seeds and seedlings, thus limiting reproduction to offshoots from established plants.

Propagation Adult plants have many suckers which are easily transplanted. They require no care but grow faster with some water and fertilizer. As is characteristic of epiphytes, this plant shows extreme efficiency of mineral utilization.

Habitat & Ecological Distribution Found on many soil types and in a broad range of dry ecological conditions. It does well even in the shade of the forest canopy and on salty sites. Direct sun and strong wind stress the plant, and very dense shade seems to stunt it.

Uses

Ornamental Grown as an ornamental for the arching leaves and attractiveflower and fruit clusters. Care should be taken to periodically remove the suckers or they will gradually form a spreading colony which will crowd out other plants. When grown in a pot it makes an attractive hardy house plant if provided with some sun.

Medicinal The juice of the young fruit contains the proteolytic enzyme bromelian which is used in Cuba to digest and eliminate intestinal parasites. The ripe fruit juice has been used in home remedies to treat fevers and ulcers of the mouth and throat and in large doses to induce abortion. Modern medicine is continuing to investigate several enzyme extracts for multiple applications.

Edible The ripe fruit may be eaten raw or cooked. It has also been used to prepare a tart beverage. The tender new leaves (*pollitas*) and flower stalks (*motate*) are sold in markets to be cooked as a vegetable.

Physical Planted as a hedge, it serves as a modest barrier to livestock and an almost impenetrable barrier to humans and dogs. Indians planted it around their villages as a barrier against surprise visitors. The leaf fibers have been used for making cloth, fishing lines, nets and as string for tying bundles.

Ecological The dense spiny patches of these plants provide refuges for many small animals such as grasshoppers, tree frogs, and ground doves. The flowers are visited by several pierid butterflies and hummingbirds.

Notes The genus is derived from the Greek *broma* meaning "food," which refers to another member of this genus, the pineapple. The species name is derived from the Latin word *pinguis* meaning "stout" or "strong." This plant is also known as wild pineapple. It is one of the few plants in which the hooked spines point in both directions. A 100 g (4 oz) sample of the tender succulent new growth suitable for use as a vegetable had the following analysis: water 92 g, calcium 158 mg, phosphorus 50 mg, iron 0.51 mg. thiamine 0.029 mg, riboflavin 0.041 mg, niacin 0.382 mg, and ascorbic acid 34 mg. Several thiol proteolytic enzymes including pinguinain and bromelian are present in the fruit and stem.

CLIFF ORCHID
Encyclia bifida   *Orchidaceae*

Geographic Distribution The genus of about 200 species is found in the American tropics. The Caribbean region has many species of these small but attractive orchids which frequently have ranges limited to only a few islands. This one is found on Hispaniola, Puerto Rico and the Virgin Islands.

Form Narrow, ridged pseudobulbs with lengthy fibrous holdfast roots lead to long narrow leathery leaves for a height of less than 30 cm (12 in).

Flower & Fruit Date The purple flowers with striped throats open serially on the end of a tall thin stalk.

Reproduction The globular seed pod with longitudinal ridges and many minute seeds is very persistent but eventually breaks open to allow dispersal of the seeds by wind. Pollination is by lepidoptera.

Propagation Heavy collecting of specimens for transplanting from the wild has greatly reduced wild populations in many areas. It is much preferred to start your own clump by acquiring a pseudobulb with a few roots from a friend. A bit of sphagnum wrapped around the base and tied in place on a rock or tree will induce the plant to grow and send out new roots. Growth rate will be more rapid if the roots have access to soil.

Habitat & Ecological Distribution

This is one of the most drought-enduring plants and may be be found growing well above the soil supported by living trees, weathered stumps, or on rocks in the full sun or open shade of the dry areas of the islands. The long stringy roots spread widely over the surface of the substrate to absorb water and nutrients from infrequent rains. If given the opportunity they will send roots delving into the soil for a more abundant supply of moisture and nutrients.

Uses

Ornamental This orchid and many other members of the genus are grown extensively on trees and rocks as part of a landscape design. They are also grown as pot plants in the Caribbean and in greenhouses in more intemperate climates.

Medicinal Many species of orchid in this genus have various alkaloids which are being tested for medicinal activity. Folk medicine has used the juice as a purgative, antihelmenthic, diuretic, against dysentery and to heal sprains and dislocations.

Ecological This orchid provides nectar for night-flying moths which are attracted by a unique combination of volatile chemicals.

Notes The genus is from the Greek in reference to the flower structure in which the labellum encircles the column. The species *bifidum* refers to the divided flower parts. This orchid has previously been known as *Epidendrum bifidum*.

AUSTRALIAN BEEFWOOD
Casuarina equisetifolia   *Casuarinaceae*

Geographic Distribution The genus has 45 species in Australia and Asia. This species has been widely planted on tropical shores of the world.

Form A tall slender tree growing to 30 m (100 ft) tall with a trunk 1 m (3 ft) in diameter. The dark-green needlelike twigs to 10 cm (5 in) in length function as leaves and are shed annually. The tiny scalelike leaves are found in whorls at the nodes of the twigs. The roots have nodules with symbiotic nitrogen-fixing bacteria allowing the tree to thrive on nutrient-poor soils. Some trees growing on waterlogged soils produce prop roots from the trunk.

Flower & Fruit Date The minuscule brown flowers, specialized for wind pollination, occur throughout the year.

Reproduction The cone is a rugose brown ball composed of paired scales which open in dry weather, releasing the seeds.

Propagation Cuttings are sometimes diffi-cult but usually root easily and grow vigorously. The potting medium should be inoculated with some soil from established plants to establish sym-biotic bacteria. Shallow planting of seeds can be effective, but application of a powdered insecticide is often needed to keep ants from carrying the seeds away. Germination takes about 10 days, yielding a seedling 15 cm (6 in) in height at 6 weeks. Growth rate is rapid with trees reaching up to 3 m in height a year after planting and 8 m after 4 years.

Habitat & Ecological Distribution The Australian pine is an aggressive pio-neer species and is very salt-tolerant. It does best on protected sandy seacoasts and may also grow inland at low elevations on a variety of soil types.

Uses

Ornamental It has frequently been planted as an ornamental, as a windbreak, and more rarely pruned into a hedge or topiary.

Medicinal The astringent bark has been used medicinally to control diarrhea and dysentery. The ground fruit mixed with powdered nutmeg has been used as a home remedy to relieve toothache.

Edible The foliage has a protein content of over 9% and is eaten by livestock when food is scarce.

Toxic The pollen often causes allergies.

Physical The fine-textured pink to dark-brown wood is hard and heavy but very susceptible to termites and should not be used in contact with the ground. It is difficult to work, splits readily and does not hold nails, but its tough elastic nature allows it to be used for wheels, rollers, tool handles, utility poles, mine props, oars and boat-building. It is an excellent firewood, burning even when green. The wood may be chemically pulped for making writing and printing paper. The bark is a source of tannins which penetrate the hide rapidly and produce a pliant reddish-brown leather. The bark is also a source of red and blue dyes. The fruits are used in various handicrafts.

Ecological It is often used to stabilize shorelines because of its ability to rapidly produce a seaside shade tree on degraded soils and deposit a "felt" of shed foliage to stop wind erosion. Growth rates in excess of 1 inch diameter per year may be reached under ideal circumstances. Casuarina was planted as a coastal shelterbelt on an area of shifting dunes on Nanshan Island, resulting in a 60% reduction in wind speed and a 12% reduction in evaporation. The stabilized sand allowed tripled agriculture yields, and the island began to export rather than import fuel and lumber. The symbiotic nitrogen-fixing nodules on the roots significantly improved the fertility of impoverished soils on which the tree was grown. On the negative side, dense shade, accumulation of shed foliage and allelopathic effects usually combine to greatly reduce or eliminate grasses and other low-growing, soil-stabilizing plants under Casuarina. Thus the tree has the potential to contribute to erosion. On many coastal and island ecosystems this tree completely overwhelms the native vegetation and alters the habitat to render it inhospitable to most species of the native fauna. In parts of Florida the planting of this tree is prohibited and its destruction is encouraged.

Notes The genus name refers to the superficial resemblance of the foliage to cassowary feathers. The species is named after the resemblance of the leaves to those of *Equisetum*. This tree is also known as Australian pine (although it is not a pine, it superficially resembles one) and ironwood. The growth of symbiotic nitrogen-fixing organisms in root nodules (as present in many legumes) helps provide the nutrients for rapid growth on poor sites. The bark contains 6 to 18% tannin composed of d-gallocatechol and pryrocatechol. The fruits contain ellagic acid, beta-sitosterol, and a kaempferol galactoside named trifolin.

SHORTLEAF FIG
Ficus citrifolia   *Moraceae*

Geographic Distribution The genus has 600 species widely distributed in the tropics. This species is found from South Florida and the Bahamas throughout the Caribbean to Paraguay.

Form A small tree, frequently bent and formed by the wind, sometimes spreading laterally as much as 40 m (130 ft) when pushed by the trade winds in exposed locations. On protected sites it may grow to 20 m (65 ft) in height. The roots may extend over 100 feet to moisture or good soil. The pointed elliptic leaves with long petioles are variable in size and shape depending on environmental conditions.

Flower & Fruit Date The minuscule flowers are borne inside a globular receptacle (the fig) which has a small pore to admit the symbiotic pollinating wasps.

Reproduction The round fleshy fruits contain many seeds and are attractive to birds and mammals when ripe.

Propagation Air layering works very well if you have access to an established tree. Cuttings treated with rooting hormone and inserted in moist soil grow readily. It is usually difficult to extricate enough root material to transplant seedlings from the wild. Seeds are also a potential source of new plants. Once started, the young trees grow rapidly.

Habitat & Ecological Distribution An extremely virile species which is often one of the few plants surviving on small rocky islands. It may grow almost any place that birds drop the seed, from a crack in the rock of a sheer cliff only slightly above the reach of waves, to high in the crotch of another tree as an epiphyte. After sending roots to the ground the fig may encompass and shade out the host tree.

Uses

Ornamental With a dense green foliage and easily varied shape, this tree has been widely used as an ornamental and shade tree throughout South Florida and the Caribbean.

Edible The fruits, reddish at maturity, are edible but tasteless.

Physical The wood is lightweight and soft but is tough and strong for its weight. It is used in guitar construction.

Ecological This tree is ideally suited to providing attractive stabilization of steep rocky hillsides.

Notes *Ficus* is derived from the Latin name for fig. The species name *citrifolia* refers to the citrus-like form of the leaves. This species has also been known as *Ficus laevigata*. Because they are hardy and root readily they are used as living fence posts. Due to their aggressive growth habit they should not be planted near foundations or water pipes. The trees can be trained into many forms including the braiding and knotting of branches, topiary and espaliered. They adapt well to bonsai treatment. The agaonid wasp *Pegoscapus assuetus* has been found to be responsible for the specialized pollinization of this fig.

SEA GRAPE
Coccoloba uvifera *Polygonaceae*

Geographic Distribution There are about 150 species of *Coccoloba* in tropical America. The sea grape is found throughout the American tropics and has been introduced in the Hawaiian Islands, the Philippines and Zanzibar.

Form A small tree with coarse branching which may reach 15 m (50 ft) in height and 30 cm (12 in) in diameter or become a prostrate shrub on tradewind-swept shorelines. The large round leathery leaves may reach 25 cm (10 in) in diameter and often have red veins. The roots tend to be shallow but abundantly extensive and intertwined.

Flower & Fruit Date The small white fragrant flowers borne on long racemes may be found in any month but mostly March to September.

Reproduction The fruits, which turn purple when mature, each have a single large seed and occur in elongate clusters. The individual berries in a cluster ripen independently of each other over a period of time.

Propagation The seeds germinate readily when planted at a depth equal to their length without being allowed to dry out. The forests of seedlings which sprout beneath prolific trees may also be transplanted. Female plants selected for fruit quality can be propagated by cuttings, air layer-ing, ground layering and grafting. This tree has a great potential for improvement by selection of certain strains for fruit quality and growth habit. A well-

44

drained site is preferred and full sun is mandatory. Although the plant is very drought-resistant, water and fertilizer greatly accelerate the generally slow growth rate.

Habitat & Ecological Distribution The salt-tolerant floating seeds are widely distributed by winds and tides. While a sandy site above the high water mark is preferred, sea grapes may be seen growing from cracks in rock with only a vestige of clay present. The tree is extremely tolerant of drought and salt but not of freezing weather. Germination of the seeds is enhanced as they are scattered in the droppings of iguanas and other wildlife which have consumed the ripe fruit.

Uses

Ornamental The large, dark-green leathery leaves with red veins make the sea grape an attractive landscape border tree. Sea grapes can be pruned into hedges, espaliered or trained into other shapes. It is long-lived but somewhat slow growing.

Medicinal The astringent roots, leaves, and bark have each been used to make a tea to treat hoarseness, asthma, hemorrhage and diarrhea. Externally, the decoctions have been applied to wounds, eruptions, rashes, and hemorrhoids. A resin collected from wounds to the bark has been marketed as "kino" and used as a source for astringent medications.

Edible The ripe fruits are edible raw and may be made into fresh or winelike alcoholic beverages. The fruit can also be made into a slightly astringent deep-red jelly.

Physical The hard, reddish, fine-grained wood takes a handsome polish and has been used for cabinetwork and furniture. It makes an ideal fuel with intense heat and little smoke. The bark of the tree contains 25% tannin. The red sap (kino) has been used for tanning and dying. Historically the leaves were used as a paper substitute.

Ecological This is one of the first woody plants to colonize sandy shores and is more tolerant of salt than most trees. Thickets of these trees can become the major factor preventing erosion of sandy shores by hurricanes. Many species of birds, deer and squirrels find the fruit very desirable. In the West Indies, hawksbill turtles frequently choose to lay their eggs in a hole they dig among the roots of beachfront sea grape trees. Ground iguanas eagerly consume the fruits and disperse the seeds. Ecological studies have shown that on small islands the presence of lizards significantly reduces insect damage to the leaves. Thus while providing good lizard habitat the tree has a symbiotic relationship in which it benefits from the presence of lizards. Buprestide beetles seem to distinctly favor the dead wood of sea grape as a place to deposit their eggs, thus leaving many larval chewed tunnels in the wood.

Notes The genus is derived from the Greek words meaning "berry" and "pod," referring to the structure of the fruit. The species name *uvifera* is derived from the Latin word meaning "grape bearing." This was probably the first land plant seen by Christopher Columbus. The flowers produce abundant nectar yielding a light amber spicy honey.

SAMPHIRE (SALT-WEED)
Blutaparon vermiculare   *Amaranthaceae*

Geographic Distribution The genus of 10 species is found worldwide in the coastal tropics. This species is found in the southern U.S., the Caribbean and the west coast of Africa.

Form A prostrate succulent annual or perennial herb with stems to 1.2 m (4 ft) long with opposite fleshy leaves. It is often found climbing on adjacent driftwood, rocks or other vegetation.

Flower & Fruit Date The rounded flower heads are 12 mm (1/2 in) or more in length on a spike and appear mostly in the spring.

Reproduction The small seeds are dark brown and lustrous.

Propagation Sections of rooted rhizomes are easily transplanted. Vegetative cuttings may also be rooted.

Habitat & Ecological Distribution It is found on saline soils, damp sand, dunes and rocky shorelines.

Uses

Ornamental This plant can be used as a contrasting ground cover in salty difficult spots.

Edible The stems and leaves may be cooked and eaten. It is sometimes fed to poultry.

Ecological The dense mats formed by colonies of this plant serve to control erosion and stop soil movement.

Notes *Blutaparon* is a corruption of the Latin *volutum laparum,* meaning loose climber. V*ermiculare* means "wormlike." The genus has also been known as *Philoxerus, Caraxeron* and *Iresine* with the common names silver head and salt-weed.

SEA-BLIGHT
Suaeda linearis   *Chenopodiaceae*

Geographic Distribution The genus of over 100 species is found worldwide. This species is found on the Gulf Coast and throughout the West Indies.

Form A small, profusely branched, slightly woody herb to 1 m (3 ft) tall. The linear alternate leaves are quite variable in size. It is a shallow-rooted annual.

Flower & Fruit Date The spike with many minute green flowers is usually borne in the summer.

Reproduction The many small seeds are smooth and shiny.

Propagation The seeds are easily collected and may be started in pots or sown directly. Water with a high calcium content is toxic to this plant unless some sodium chloride is also present.

Habitat & Ecological Distribution This plant is usually found on slightly elevated drier saline soils adjacent to mangroves, marshes, or salt flats. Experiments have shown that it is one of the few plants that grows better with some salt in its environment even in the absence of competitors.

Uses

Ornamental It is a useful ground cover for saline soils.

Edible The young leaves and stem tips may be cooked and eaten as a vegetable. A change of water reduces the saltiness, or a small amount may be used to provide the salt in soups and stews.

Ecological This plant is often one of the pioneers to vegetate and stabilize harsh or disturbed sites. The seeds are eaten by birds, mice and land crabs. Aluminum and copper compounds in the soil frequently inhibit growth of this plant and others. When growing in a salty soil, *Suaeda* growth is stimulated by low levels of these metals.

Notes This plant is also known as *S. torreyana* and *S. conferta*. The foliage has 2% protein, 0.4% fat and significant amounts of calcium, phosphorus, iron, thiamine, riboflavin, niacin and vitamin C. The betacyanin pigments in the leaves are composed primarily of celosian with lesser amounts of suaedin and amaranthin.

GLASSWORT

Salicornia bigelovii   *Chenopodiaceae*

Geographic Distribution The genus has about 35 species distributed worldwide in the salty tropics. This annual species is found from Nova Scotia to California, Mexico and the West Indies. The similar perennial species *S. virginica* is found from Alaska to California, Massachusetts to the West Indies, and Europe to North Africa.

Form A fleshy herb composed of naked translucent cylindrical branched spires with no obvious leaves or flowers, growing 12 to 50 cm (5 to 20 in) tall, green initially, but often turning red at upper parts at maturity. The leaves are present in the form of small scales at the nodes. S. *virginica* has trailing, often forked, decumbent stems rooting at the nodes. The many erect branches are green turning to brown or lead color with age.

Flower & Fruit Date The minute inconspicuous flowers occur in threes at the nodes.

Reproduction Extensive colonies are formed from the extended branching rhizomes. The tiny hairy seeds are widely dispersed by water.

Propagation Seeds of *Salicornia* seem to require a month or more of post-maturity ripening to reach high percentages of germination. They thrive in areas of very high soil saltiness with germination proceeding vigorously in salinities up to twice that of the open sea. Seedling establishment is inhibited if the newly germinated seeds are watered with fresh water. Newly germinated seedlings require several days free of tidal submersion; thus they can only propagate naturally in many areas in neap tide periods. If transplanted, *Salicornia* will often thrive and spread vegetatively as mature plants in spots that it has not naturally invaded as seedlings.

Habitat & Ecological Distribution In the moist upper areas of marshes and on sandy, rock, or clay shores, most commonly found on intermittently submerged bedrock in shallow protected areas. *Salicornia* is one of the best known salt-requiring plants which show optimum growth in saline media. It does not tolerate long inundation but continues to thrive under hypersaline conditions. Flooding of the habitat by fresh water results in rapid replacement by other species.

Uses

Ornamental The attractive red and green branches make this plant a candidate for planting in an otherwise inhospitable salty damp depression.

Edible All the species can be cooked fresh, pickled or preserved and are considered a delicacy. They may also be used to provide a tart salty crunch in salads. It is lightly grazed by livestock including goats, donkeys, and camels.

Physical When under cultivation and irrigated by seawater an annual productivity of 20 tons per hectare has been obtained.

Ecological This is one of the few low ground covers which will grow in saturated hypersaline soils.

Notes The genus name *Salicornia* is derived from the Latin words meaning "salt" and "horn." The species *virginica* refers to its being originally found in Virginia. The species *bigelovii* honors Jacob Bigelow, an American botanist. Another common name is wild coral. The foliage contains: 5.7% protein, 0.4% fat, 17% fiber, 41.3% ash, 0.1% phosphorus, and 1.5% oxalate. Several new commercial strains of *Salicornia* have been developed which can grow on soil that is irrigated with seawater. A hectare of one type will support up to 20 goats with a weight gain similar to that of a hay diet. A second type selected for its oily seeds can produce up to 3 tons per hectare of seeds which yield 30% oil; the byproduct residual meal contains over 40% protein and can then be fed to livestock and poultry. Betacyanin pigments isolated from the foliage have been found to be primarily celosianin with about 8% amaranthin.

BLACK MAMPOO
Guapira fragrans   *Nyctaginaceae*

Geographic Distribution The genus has 50 species in the American tropics. This species occurs in the Florida Keys and throughout the Caribbean but not in the Bahamas.

Form A medium-sized evergreen tree with smoothish gray bark and an open spreading crown, growing to 12 m (40 ft) tall with a trunk 60 cm (24 in) in diameter.

Flower & Fruit Date Male and female flowers are on different trees in the early spring, each sex borne terminally on twigs and composed of branched clusters of small inconspicuous greenish flowers.

Reproduction The oval, single-seeded fruits are eaten by birds and thus broadly dispersed.

Propagation The tree grows well from seed, even when deposited by bird droppings.

Habitat & Ecological Distribution
A salt-tolerant tree which grows from the seacoast to over 300 m (1000 ft) in elevation particularly on sites that have more moisture or are protected from the tradewinds.

Uses

Ornamental While it forms an interesting shade tree when established, its roots tend to be aggressive.

Medicinal A tea from the leaves has been used as a home remedy to treat typhoid fever.

Physical The soft, punky, tenacious wood is sometimes used for slack cooperage, rough boxes or a low-grade fuel whenother timber is scarce.

Ecological The fruits are eagerly sought by many species of birds.

Notes The genus is derived from an Indian name for the tree. The species refers to the abundant fragrant flowers. This tree has also been referred to as *Torrubia* and *Pisonia*.

WATER MAMPOO
Pisonia subcordata   *Nyctaginaceae*

Geographic Distribution The genus of about 50 species is found worldwide in the tropics. This species is found from Puerto Rico south through the Lesser Antilles.

Form A sturdy tree to 12 m (40 ft) tall with a rounded to widely spread crown and a gray trunk to 1 m (3 ft) in diameter. The rounded opposite leaves are slightly shiny dark green above and lighter below.

Flower & Fruit Date The hairy greenish male and female flowers are in ball-like clusters on stalks on separate trees. The flowers are fragrant and entire trees hum with bees when in bloom.

Reproduction The entire clusters of fruits drop from the tree when ripe. Each fruit contains a single seed enclosed in a cylinder with five rows of sticky glands.

Propagation The tree grows well from seeds and cuttings.

Habitat & Ecological Distribution This tree is very tolerant of salt and poor conditions. It is found on many soil types in a coastal dry forest habitat from wind-battered sea cliffs to inland hills at low elevations.

Uses

Ornamental It is often retained as an ornamental when wild land is cleared because of its pleasing form and the fact that it is one of the few trees which grow to a large size on poor dry soils.

Physical The soft white wood is lightweight, porous and favored by termites. It has been used for net floats, construction of packing crates and fuel.

Ecological The sticky fruits are widely dispersed by adhering to animals.

Notes The genus is named for William Piso, a 17th-century physician who traveled and wrote about Brazil.

SALTWORT

Batis maritima   *Bataceae*

Geographic Distribution This and a similar Australian species are the only members of the family. It is widespread in the American tropics and has been introduced and naturalized in the Hawaiian islands.

Form A yellowish-green, prostrate, pungent shrub with branches to 1.2 m (4 ft) long, rooting at soil contact and spreading further to form extensive colonies.

Flower & Fruit Date The minute flowers are found throughout the year at the base of the leaves.

Reproduction The abundant 1-cm (1/2-in) fruits look like miniature potatoes but are actually sturdy berries.

Propagation Well-established rooted branches may be directly transplanted. Cuttings or rooted runners are best started in coarse silica sand. Seed may be started in a slightly salty medium.

Habitat & Ecological Distribution As the name suggests, saltwort is extremely tolerant of saline saturated soils and may be found bordering salt ponds, marshes, salt flats and fringes of mangrove mud.

Uses

Ornamental An excellent low-maintenance ground cover for moist salty places.

Medicinal As a home remedy, a tea prepared from the tender parts has been used to treat venereal diseases, skin diseases and asthma. Folklore states that daily consumption of the raw leaves is a remedy for constipation, rheumatism and gout. The finely ground leaves have been used as a poultice for wounds.

Edible The leaves may be eaten raw, boiled and strained for a vegetable puree, or pickled. Pouring off the first cooking water eliminates much of the excess salt.

Physical An annual production of 17 tons per hectare has been obtained with seawater irrigation.

Ecological Many herbivo-

rous animals include small amounts of this plant in their diet when it is available.

Notes The genus name *Batis* is derived from the Greek, making reference to the fruit resembling a blackberry. The species *maritima* is from the Latin referring to its growth near the sea. The flavonoid pigment in the leaves has been identified as isorhamnetin 3-0-rutinoside.

SEA PURSLANE
*Sesuvium   portulacastrum
Aizoaceae*

Geographic Distribution The genus of 10 species is found worldwide in the tropics. This species is ubiquitous on tropical and subtropical shores worldwide.

Form A fleshy prostrate perennial herb which sprawls and branches across the substrate, often sending down roots at the nodes. The thick fleshy leaves are up to 5 cm (2 in) long by 1 cm (1/2 in) wide. They are usually dark green but may be tinged with or entirely red.

Flower & Fruit Date The small white to purple flowers occur in the leaf axils throughout the year.

Reproduction The fruit is a small capsule which splits open around the axis to release the shiny black seeds. The part of the capsule remaining on the plant is reminiscent of an egg cup.

Propagation Rooted cuttings are the best method to produce large numbers of plants. Sea purslane seeds are difficult to collect but germinate and grow well when planted in salty moist sand. The plant also responds well as transplanted rooted nodes. Excessive irrigation often results in a fungus infection.

Habitat & Ecological Distribution A rapid-growing early colonizer of sand and gravel beaches, it also may be found on salt flats and rocky shorelines, sometimes climbing over adjacent plants.

Uses

Ornamental It has been used as an attractive ground cover when its competitors have been excluded by hopelessly salty soil.

Medicinal This plant has been used to treat scurvy (probably effectively) and the seeds have been used to eliminate intestinal worms. External application of a leaf slurry is reported to relieve the stings of venomous fish.

Edible The crisp salty leaves make an interesting addition to a salad, or may be cooked in several changes of water to reduce the saltiness and served as a vegetable. It is sold in the vegetable markets in the Orient. Sheep, goats, burros and camels all graze on it.

Physical Formerly the ash from burning of this plant was used in the manufacture of glass and soap.

Ecological The propensity to form extensive mats makes sea purslane an

excellent binder of sand against movement by wind and waves. Severe storms may eliminate entire populations but natural regeneration from rhizome fragments and seeds buried in the sand soon results in reestablishment of colonies. Land crabs living in burrows near the sea eat it.

Notes The genus is a very old name for this plant. The species name refers to the resemblance of this plant to some of the *portulacas*. This plant is rich in ecdysterone, a rare insect molting hormone, the similar alpha-ecdysone and the flavonol glycoside sesuvin. The chemical analysis of 100 grams of the leaves shows the following composition: water 87.8, protein 2.1, fat 0.4, fiber 0.8, carbohydrate 5.8, ash 3.9, calcium 0.67, phosphorus 0.23, iron 0.12, sodium 1.12, potassium 0.32. The vitamins A, C, niacin, riboflavin, and thiamine are also present in significant amounts.

COMMON PURSLANE
Portulaca oleracea   *Portulacaceae*

Geographic Distribution The genus has about 200 species found world-wide in the tropics and adjacent areas. Native to the Old World tropics, this species is now intro-duced to warm areas throughout the world and has been found as far north as 51° in America and 54° in Europe.

Form A fleshy herb spreading red fleshy prostrate stems radially 0.3 m (1 ft) or more from a central tap root. In the shade it tends to grow more upright with fewer leaves and flowers.

Flower & Fruit Date The small yellow flowers with no stalks have 5 petals and are found throughout the warm parts of the year. The flowers open only once for about 4 hours in the middle of a hot bright day. They seem capable of self-pollination as flowers do not attract insects and those which do not open due to cloudy weather still produce pods with viable seeds.

Reproduction The fruit is a round capsule maturing in 13 days. The capsule splits open around the middle and the top half falls away like a lid releasing an average of 40 small black seeds. The seeds retain their viability for as long as 40 years.

Propagation *Portulaca* can be grown in a light potting soil starting with cuttings. Seeds require sunlight for germination and should be sprinkled on the surface of a moist sandy soil. After one year of storage or resting buried in the soil, a brief period of exposure to light will induce germination. Germination percentage and speed are both improved if the soil temperature is maintained above 25°C.

Habitat & Ecological Distribution An opportunistic plant with a wide tolerance for adversity, it is adapted to wait in the soil until a disturbance produces a clearing which can be exploited. It seems to do best on moist sandy soil but may be found on almost any substrate. It is very tolerant of salt and drought. Dispersal is primarily by animals as about 62% of the seeds are capable of germination after passing through the gut of a bird and a higher percentage of germination results from passing through livestock. The seeds can survive winters with mild soil freezing. Because of its vigorous and adaptable growth habits it is considered to be a weed when growing with many agricultural crops.

Uses

Ornamental This and other forms of *Portulaca* are used decoratively alone, as ground covers and as surface layers in pots with larger plants.

Medicinal Prepared as a tea, it has been used as a folk remedy to treat worms, as a diuretic, and to reduce fevers. The crushed leaves and juice of the plant have been used externally as an emollient, to treat prickly heat, inflammation and tumors and to heal wounds. The entire plant has also been used effectively to treat scurvy and in folk medicine to treat diseases of the liver, bladder, kidneys and lungs. Regular consumption has been recommended to overcome chronic constipation. The seeds are reputed to be a diuretic and a potent emenagogue. The juice of the pounded leaves has been consumed to treat asthma. An extract of this plant contains levarterenol, which raises the blood pressure while lowering the heart rate due to stronger shorter contractions. The medicinal value is officially recognized in the French, Spanish, Mexican, and Venezuelan pharmacopoeias.

Edible The tender young stems are eaten in salads or may be cooked and served like spinach. They may also be mixed with crumbs and beaten egg before being fried or baked. The seeds may be ground into an edible meal. It is eaten by livestock and in many areas it is considered to be an excellent pig food.

Toxic In some soils an excessive accumulation of nitrates renders this plant somewhat toxic to cattle. The oxalic acid content of the leaves varies from 3.5% to 9.3% and has produced oxalate poisoning in sheep when consumed in large quantities. Tests of the leaf, fruit, stem, and root show small amounts of hydrocyanic acid. Humans are advised not to regularly eat large amounts of this plant.

Ecological As it is often one of the early pioneers, it helps stabilize disturbed sand.

Notes *Portulaca* is an ancient Latin name for this plant and *oleracea* refers to it being like a garden vegetable. A 100 g (4 oz) sample of the foliage contains: 92 g water, 21 calories, 1.7 g protein, 0.4 g fat, 3.8 g carbohydrate, 0.9 g fiber, 103 mg calcium, 39 mg phosphorous, 3.5 mg iron, 2500 mg sodium, 0.34 mg thiamine, 0.10 mg riboflavin, 0.54 mg niacin, and 700 mg ascorbic acid (vitamin C). Also present are variable amounts of oxalic acid, cyanogenic acid, urea, alkaloids, glucoside, mucoid compounds, and potassium salts. The leaves contain the free amino acids isoleusine, aminobutyric acid and tyrosine. Two red-violet pigments have been found to be acylated betacyanins. The seeds yield 17% oil which contains beta-sitosterol.

HAIRY PORTULACA
Portulaca pilosa   *Portulacaceae*

Geographic Distribution This species is found along the coast of the Gulf of Mexico, the West Indies, Central and South America.

Form A succulent herb which may grow flat on the substrate or with upcurving branches. It is characterized by white hairs in the leaf axils.

Flower & Fruit Date The rose purple flowers surrounded by white or brown hairs about 1 cm (3/8 in) across are borne continuously throughout the year. The flowers last only about 6 hours in the middle of a hot bright day. They seem capable of self-pollination because flowers which do not open due to cloudy weather still produce pods with abundant viable seeds.

Reproduction The fruit requires 13 to 17 days to mature, then the spherical seed capsule splits open around its axis, releasing about 60 small, black, long-lived seeds.

Propagation Cuttings root well and may be used to start new plants. Seeds require sunlight for germination and should be sprinkled on the surface of a moist sandy soil. Germination percentage and speed are both improved if the soil temperature is maintained above 25°C.

Habitat & Ecological Distribution This portulaca may be found on sandy and rocky shorelines as well as inland as a pioneer species on open sites. Dispersal is aided by wildlife as about 16% of the seeds are capable of germination after passing through the gut of a bird. Most of the seeds are killed by freezing temperatures.

Uses

Ornamental It may be used as a border or an attractive carpetlike ground cover under more erect plants.

Medicinal The juice has been used to treat skin infections and a tea has been used to treat worms, scurvy and sleeplessness. It has also been used as a quinine substitute (probably unsuccessfully) to treat malaria.

Edible The young plant may be cooked and eaten.

Toxic A dark-green oil extracted from the seeds is toxic, but in small doses it expels intestinal worms.

Physical The entire plant has been used as a hair pomade.

Ecological This is one of the large community of inconspicuous but important binders of sand and soil along the seashore.

Notes The species name *pilosa* refers to the hairs found on the stem. It has been introduced and become naturalized in Southeast Asia. This herb contains the free amino acids argenine, cystine, alanine and methionine as well as the chro-moalkaloid betacyanogen.

SEA ROCKET
Cakile lanceolata   *Brassicaceae*

Geographic Distribution The genus of 15 species is found bordering oceans and large salt lakes in temperate and tropical North America, Eurasia and Australia. This species is found in South Florida, Mexico and the West Indies.

Form A fleshy spreading herb to 60 cm (2 ft) tall, often forming extensive clumps. The leaves are alternate oblong and coarsely toothed.

Flower & Fruit Date The small inconspicuous white flowers occur in clusters near branch tips.

Reproduction Each of the two sections of the elongate rocket shaped corky pods contains a light brown seed about 9 mm (3/8 in) long. The upper stage of the rocket is easily detached and is widely dispersed by winds and currents. The lower stage is more firmly attached to the parent plant and is likely to be the founder of new plants close to the parent.

Propagation The fresh seeds are almost 100% viable. The seeds should be planted at least 2 cm deep and the soil moisture should be carefully maintained until emergence.

Habitat & Ecological Distribution Found on the strand line of sandy beaches, contributing to primary dune formation, on the edges of saline ponds and as a marsh border.

Uses

Ornamental It makes a sand-stabilizing ground cover and attractive upland border to a beach.

Medicinal Due to the high vitamin C content this plant has been used successfully to treat scurvy. It has also been used to treat lymphatic and skin diseases.

Edible The leaves, tender stems, flower buds and green pods may be boiled for 10 minutes to make a nice vegetable with the mustard flavor characteristic of this family. Raw young shoots and seedpods make a tasty addition to a salad.

Ecological This plant is often a pioneer species which stabilizes shorelines sufficiently for the establishment of more permanent species.

Notes The genus is from the Arabic name for the plant. The species is from the Latin word meaning "tapering to a point."

LIMBER CAPER
Capparis flexuosa   *Capparaceae*

Geographic Distribution The genus of about 250 species is found throughout the warm parts of the world. The limber caper is found from South Florida throughout the Caribbean to Argentina and Peru.

Form A small evergreen shrub which may develop a vinelike growth pattern in the shade or become a small tree to 6 m (20 ft) tall with a 12 cm (5 in) trunk when growing in the sun. The alternate leaves are dull and leathery.

Flower & Fruit Date The distinctive flowers with four white petals and a dense spray of white stamens may be found year-round. The flowers release their fragrance in the evening.

Reproduction The greenish pod with a longitudinal maroon stripe opens to reveal a deep reddish interior with pearly seeds.

Propagation Limber caper germinates and grows well from seed which is picked fresh and planted immediately.

Habitat & Ecological Distribution It is widely distributed in low altitude forests. While it is moderately tolerant of salt and wind it does best if it is not in direct contact with seawater and has some shade.

Uses

Ornamental It has been trained as an attractive barrier hedge which seldom needs trimming.

Medicinal A tea made from the roots or bark is taken as a diuretic, and a tea from the leaves is applied to skin diseases. The seed pod is reputed to be an anti-spasmodic.

Toxic The juice from theroots has a pungence similar to horseradish and is reported to blister the skin.

Physical The light-brown, hard wood makes resilient tool handles, and the frayed twigs are used as toothbrushes.

Ecological The "shaving brush" flowers are attractive to nocturnal nectar-feeding moths and bats.

Notes The genus name is derived from the ancient name for the European tree which produces edible capers. The species is from the Latin meaning "flexible." The inner bark and roots have a pungent penetrating smell similar to horseradish. The leaves have an interesting biochemistry with at least 5 different C4 glucosinolates, one of which has been identified as gluconorcappasalin. An isothiocyanate-producing glucoside has been identified as glucocapparin.

JAMAICA CAPER
Capparis cynophallophora   *Capparaceae*

Geographic Distribution The natural range of this species includes South Florida, the Caribbean and Central America.

Form A small tree with a rounded compact crown to 6 m (20 ft) tall and a trunk of 15 cm (6 in). The narrow alternate leathery leaves are shiny on top and dull rusty bronze below.

Flower & Fruit Date The white flowers with 5 cm (2 in) stamens are borne in clusters in any month. They open at dusk, releasing a fragrance which attracts moths. The sun of the following day causes them to gradually turn pink then a darker purple.

Reproduction The long light-brown pods with constrictions between the seeds split open at maturity to show the red interiors and allow the shiny brown seeds to fall to the ground.

Propagation Fresh seeds germinate readily and the seedlings are easily transplanted if they are kept wet until replanted. The growth rate is accelerated with a balanced fertilizer. Cuttings may also be used to start new plants.

Habitat & Ecological Distribution Its tolerance for drought and salt allow it to thrive in dry coastal forests and adjacent to mangroves. It does well as an understory plant growing in the shade of larger trees. It is sensitive to freezing temperatures.

Uses

Ornamental The hardy nature of the Jamaica caper and its attractive dense crown has resulted in its use as a park and roadside tree by landscapers. If planted close together in a row they make a tall barrier hedge which never needs pruning.

59

Medicinal The roots and seeds have been used to treat intestinal worms. A tea from the roots is used as a menstrual stimulant. A leaf tea is used as a diuretic and to treat venereal diseases.

Edible The unopened flower buds of the related *C. spinosa* are the pickled capers of commerce.

Physical The hard, heavy, fine-textured wood is yellow with a reddish tinge but has seldom been used for other than posts or fuel.

Ecological This is an ideal tree to serve as a windbreak and to help stabilize windward shores.

Notes The genus name is derived from the ancient name for the European tree which produces edible capers. The species name refers to the long phallus-shaped pods.

Cocoplum

Chrysobalanus icaco   *Chrysobalanaceae*

Geographic Distribution The genus has a species in Africa and one in America. The cocoplum is found naturally on seacoasts throughout the tropical Atlantic. It has been introduced and naturalized in the Seychelles, Vietnam and Fiji.

Form A large native shrub foliaged from the ground up. It may grow to a height of 8 m (25 ft) with a dense crown and a trunk 10 cm (4 in) in diameter on protected or favorable sites. But on windy sites or in poor growing conditions it usually forms dense thickets only a few feet high. The dark-green, leathery, elliptic leaves, yellowish on the bottom, are alternate in two rows oriented along the twig.

Flower & Fruit Date The flowers are small and white in clusters near twig ends in all months.

Reproduction The coastal forms usually have spherical white, pink or red fruit. The interior variety typically has olive-shaped fruit which is dark purple when ripe. All have a thin spongy flesh with a single nutlike seed. Cocoplums usually occur in thickets, spreading both by seeds and vegetative means.

Propagation Cocoplums grow slowly from seed although germination is hastened if the seed is cracked before planting. Seedlings or parts of an established clump may be transplanted. A better propagation method, if material is available, is to air layer or to root hardwood cuttings under mist. The form with red-tipped young leaves has been increasingly in demand as an ornamental and is propagated vegetatively. The form with white fruit is more tolerant of drought and salt. They may need extra water on dry sites until established.

Habitat & Ecological Distribution Tolerant of salt and wind, it may be found on shallow wet soils, sandy beaches, or heavy clay inland to an elevation of 500 m (1500 ft). This plant is very cold-sensitive, and a slight touch of frost is usually lethal. It needs full sun to thrive.

Uses

Ornamental They make excellent landscape plants with a densely foliaged symmetrical growth pattern, dark evergreen leaves, freedom from pests and disease, and positive response to hedging and pruning. The light fruited form is more salt and wind resistant and should be selected for planting near the shore. Both forms thrive on more inland sites.

Medicinal Teas from the bark, fruits, roots, and leaves have been used variously in home remedies to treat diarrhea, dysentery, hemorrhages, and as a laxative. The kernel oil has been used as a base in medical ointments.

Edible The slightly astringent sweet fruit and the kernels in the seed are edible raw or cooked. The fruit may be dried or canned as a sweet preserve or bright red jelly.

Physical The hard, reddish-brown, heavy, durable wood has been used in carpentry. The leaves and fruits have been used as a source for a black dye. Tannin from the fruits produces a soft and porous leather. The kernel has been burned by Indians for illumination.

Ecological The low bushy growth habit makes cocoplum a very effective plant for stabilization of sand dunes. Many species of wildlife enjoy the flesh and seeds of the fruit.

Notes The genus is from the Greek word meaning "golden acorn," referring to the golden hue of the fruits of some varieties. The species name is derived from the Arawak Indian name for this plant. This shrub is also known as *icaco* or fat pork. The nectar produces a dark rich honey. A 100 g (4 oz) sample of the fruit contained 14 g sugar, 23 mg calcium, 17 mg phosphorus, 0.81 mg iron, 0.047 mg thiamine, 0.53 mg riboflavin, 0.42 mg niacin, and 10 mg vitamin C. The kernels may be over 50% oil and have a potential commercial value. The fatty acid composition of the oil is: eleostearic 22%, stearic 18%, oxyparinaric 18%, oleic 11%, parinaric 10%, licanic 10%, linoleic 6%, palmitic 4%, and arachidic 1%.

Nicker bean
Caesalpinia bonduc   *Caesalpinaceae*

Geographic Distribution This genus has 200 species of thorny shrubs and trees in the tropics worldwide. This species is found worldwide on tropical shores.

Form A straggling vinelike shrub to 8 m (25 ft) long with many curved spines, often climbing on adjoining plants. The bipinnate leaves also have hooked spines.

Flower & Fruit Date The small greenish-yellow flowers are borne on erect stalks in the winter or spring.

Reproduction The broad seed pods, densely covered with sharp spines, open to expose the slightly oval blue-grey seeds. The closely related *C. ciliata* has yellow-orange seeds.

Propagation The extremely tough seed coats should be scarified by filing or grinding, otherwise germination of the seeds may take several years. Planting of the mature seeds while they are still green also hastens germination.

Habitat & Ecological Distribution The nicker bean is very tolerant of salt and drought and thrives on sand dunes or coral rubble berms as well as rocky clay upland soils.

Uses

Ornamental As a tough thorny vine the nicker bean can be trained as an attractive and unobtrusive but impenetrable barrier to people and livestock.

Medicinal Bonducin obtained from the seeds or bark as a tea has been called "poor man's quinine" and has been used to treat intermittent fevers. It has been claimed to be as effective as quinine in treating malaria but other experiments have found it to be ineffective in treating avianmalaria. The crushed seeds have been used to make a tea to treat hemorrhoids, kidney problems, venereal diseases, hypertension and diabetes. Use by diabetics is said to be dangerous because the seeds suppress urinary sugar while not affecting blood sugar. The ground and roasted seeds have been given as a diuretic to treat cardiac or renal edema. The new leaves have been eaten to treat menstrual disorders and expel worms and have been applied as a poultice to toothaches. The seed oil has been used externally to treat arthritis and as an emollient in otitis. Modern medicine has found various extracts to be antipyretic, diuretic, anthelmintic, antiviral, and active against sarcoma.

Edible The crushed roasted seeds have been used to make a coffee substitute.

Toxic The unroasted seeds are poisonous. An oral dose of 4 mg per kg of the powdered seeds has been shown to effectively eliminate intestinal worms in water buffalo calves. The seeds are reported to be effective in evicting unwanted burrowing land crabs (*Cardisoma guanhumii*) which are inclined to consume all plant sprouts near their burrows. Apparently when the seed is dropped down a burrow the crab has a very frustrating time trying to grasp the hard smooth seed for removal from its burrow and eventually gives up and moves to a new location.

Physical Children use the seeds as marbles, or rub them on abrasive surfaces until they are hot enough to "scorch" their playmates. The seeds are also used in various handicrafts as beads and often as the playing pieces in Wari, a traditional African game widely played in the West Indies. The oil from the seeds has been used in cosmetics and medicinal preparations. The thorny stems have been used in Polynesia to snag fruit bats. The wood, known commercially as Pernambuco, is hard, compact and strong with a rich red color, but is seldom available for items larger than inlays or small carvings.

Ecological The seeds remain viable for long periods of time while being transported by ocean currents, thus this plant has become widely established, even on remote small islands.

Notes The names *Poinciana* and *Guilandina* have also been used for this genus. *Caesalpinia bonduc* has also been called *C. crista* and *C. ciliata* has been called *C. major*. The glycoside bonducin used as a quinine substitute is an amorphous white bitter powder with the formula $C_{20}H_{28}O_8$. This chemical has also been called guilandinin. The seeds also contain the homoisoflavonoid bonducellin, saponin, sugars, starch, beta-sitosterol, heptacosane, furanoditerpenes, the enzymes protease, urease, amylase, peridoxidase, catalase, oil and 3 related bitter compounds called caesalpins. The oil contains palmitic, stearic, lignoceric, oleic, and linoleic fatty acids. The wood contains 2 dyes, brasilin $C_{16}H_{14}O_5$ and brasilien $C_{16}H_{12}O_5$ which are related to haematoxylin. These dyes produce a bright scarlet color when applied to cloth and leather but are prone to fade in the sun. These dyes were also used in staining tissue sections for histological examination. Fossil *Caesalpinea* have been found in Tertiary rocks.

Pride of Barbados
Caesalpinia pulcherrima *Caesalpinaceae*

Geographic Distribution Native of the New World tropics (or Madagascar), this species is now widely introduced and naturalized throughout the warm parts of the world.

Form A prickly open-crowned shrub to 5 m (15 ft) in height with large fine bipinnate leaves.

Flower & Fruit Date The showy brilliant flowers are present almost continuously. All three flower colors (red, yellow and orange) may appear on the same plant, but always in separate inflorescenses. It may flower as soon as 8 months after planting.

Reproduction The flat dark-brown pods contain 5 to 8 shiny brown 9-mm (3/8-in) beanlike seeds.

Propagation If the seeds are soaked for 2 days prior to planting they will germinate readily and grow rapidly.

Habitat & Ecological Distribution This hardy leguminous salt-tolerant plant thrives undaunted in full sun on poor dry soils of many types. It is very sensitive to frost, but can stand prolonged flooding of up to six weeks.

Uses

Ornamental Generally it is planted as a border, screen or unclipped hedge. It makes a bright accent among other plants in a small setting.

Medicinal The flowers have been prepared in a tea to treat menstrual problems, induce abortions and as a purgative. Tea from the yellow-flowered form is used as a gargle for sore throats and when beaten with an egg to relieve coughs. Tea from the red-flowered form is taken to relieve cold symptoms and digestive disorders. Leaf tea has been used to treat skin diseases, ulcers of the mouth and throat, and liver afflictions. The bark and roots have also been used medicinally and to induce abortions. Extracts of the plant have experimentally been found to protect chick embryos against vaccinia and influenza virus. Antibiotic activity has also been recorded against gram negative and Candida bacteria.

Edible The immature seeds are edible. A trypsin/chymotrypsin inhibitor has been isolated from the seeds. These protease inhibitors reduce the nutritional value of the seeds to potential consumers.

Toxic The crushed leaves yield hydrocyanic acid and have been used as an intoxicant to capture fish. The dried leaves fed to rabbits caused their death several days later. Extracts of the flowers have been found to have molluscicidal properties against disease-carrying snails.

Physical A red dye has been prepared from the roots. The seed pod is a source of a black or yellow dye and has been used for tanning hides. The dried powdered flowers may be used as an insecticide.

Ecological As with many of the legumes, symbiotic root nodules help the plant by fixing atmospheric nitrogen into a usable form. This process also improves the general fertility level of the soil.

Notes The genus is named for Andreas Caesalpini, a 16th-century Italian botanist who published one of the first plant books of the Renaissance. He based his taxonomy on the form of reproductive structures although most theologians of the day denounced the concept of sex in plants. Perhaps being chief physician to the Pope allowed the blasphemous ideas which are the basis of modern botanical taxonomy. The species name is from the Latin word meaning "beautiful." The genus *Poinciana pulcherrima* has been used for this plant and the common name flower fence is also used. The leaves contain 7% tannin, gallic acid, gum, resin, bonducellin, benzoic acid, peltogynoids, homoisoflavonoids, caesalpins, a benzoquinone and a chalcone. A new cassane-type diterpene ester named pulcherralpin; the peltogynoids, pulcherrimin and 6-methoxypulcherimin and the homoisoflavonoids bonducellin and 8-methoxybonducellin have been isolated and characterized from the stem. The stem bark contains ellagic acid, leucodelphinidin, gallic acid, beta-sitosterol, sebacic acid, quercimeritrin, leucoanthocyanin, two ellagitannins and other tannins. The root contains a series of furanoditerpenoids. The flower pigments include the anthocyanin cyanin. The seed contains about 31% of a galactomannose polysaccharide mucilage and the only known glycoprotein protease inhibitor. The flowers yield a high-quality honey.

Tamarind

Tamarindus indica   *Caesalpinaceae*

Geographic Distribution The tamarind is the only species in the genus and is native to tropical Africa. Its early introduction and naturalization in India led to the common and generic names being derived from the Arabic *tamr hindi* meaning Indian date. The species name also refers to an Indian origin. The fruit was an article of commerce in the Mediterranean as early as the 4th century B.C. It is now introduced and naturalized throughout the tropics.

Form A very slow-growing noble tree with a potential of 30 m (100 ft) in height and a trunk 4 m (13 ft) in diameter. The dense rounded crown has fine bipinnately compound leaves.

Flower & Fruit Date In the fall the racemes of bright pink flower buds give rise to pale yellow flowers with orange to red streaks.

Reproduction The thick cinnamon-brown pods have a brittle outer shell enclosing the shiny squarish seeds, which are embedded in a sweet but astringent pasty pulp held together by strands of fiber. The pods ripen in late spring about 10 months after flowering. The tree usually bears its first crop of pods after 7 to 10 years. The ripe fruit is subject to attack by mold, insects and birds.

Propagation The tree grows readily from seed but is slow to germinate unless the seeds are soaked in water for several days. The seeds remain viable for long periods of storage and sprout easily after passing through the gut of livestock. Initial growth rate is rapid with tap root development of 30 cm (1 ft) possible in the first 60 days. Growth rate is slower with increasing age. There is considerable variation in quality and quantity of fruit. Trees with the most desirable fruit characters can be grafted or air layered and will produce fruit at a younger age. In plantations, tamarind should be spaced 8 to 12 m (27 to 40 ft) between trees.

Habitat & Ecological Distribution The tamarind is salt- and drought-tolerant and grows well on many types of soils near the sea. It is one of the few trees that will thrive and produce dense shade even on dry windward coasts. The strength of the branches results in a wind tolerance which even extends to hurricanes. It requires dry weather for successful fruit development and is very sensitive to frost. The tree is not tolerant of shade and does not regenerate under the dense shade of its own canopy.

Uses

Ornamental A very attractive tree when mature. Its shade is often the locus for informal social gatherings.

Medicinal The fruit has been used medicinally for at least 3000 years. The ripe pulp is widely used as a laxative, diuretic, antiscorbutic and to treat fevers. Other folk uses include the treatment of indigestion, bile disorders, sore throat, sunstroke, alcohol intoxication and the restoration of sensation after paralysis. The high vitamin C content of the pulp makes it an effective treatment for scurvy. A tea prepared from the leaves has been used to treat worms, dysentery, diabetes, coughs, as a diuretic and as an eyewash for infected or inflamed eyes. Strips of young bark have been pounded, cooked and eaten to cure diarrhea. A tea from the roots has been used to treat constipation, chest congestion and leprosy. The seed cut in two and rubbed on a scorpion sting is said to be a certain cure. Modern medicine has determined that the ripe fruit has antibiotic activity against gram positive and negative bacteria, yeast and fungi.

Edible The young pods contain little of the sugar or acid present in the ripe fruit and may be used fresh as a seasoning in cooked dishes or preserved in syrup. The young leaves which are high in protein are used as an ingredient in soups and salads and some curries. The ripe fruit has several outstanding characteristics and is so appreciated that 500,000 tons per year are harvested in India alone. It is the fruit with the lowest water content and is the most acidic with a high cream of tartar content. It is equalled in sugar content only by dates. The pungent pulp of the ripe fruit is often separated by hand and sold in compressed cakes. The fruit pulp is also easily separated as a slurry by boiling for a few minutes with enough water to cover it. The resultant syrup may be made into various beverages or used as a sauce on meat. The pulp is used as a flavoring agent in curries, chutney, and in worcestershire and proprietary sauces as well as being a traditional ingredient in southern Indian cooking. The pulp may be mixed with sugar as a confection, swirled into ice cream or used to make an excellent tart jam. The seed kernel may be eaten after being roasted or boiled, or it may be ground into a flour and mixed with other cereals for the preparation of bread and cakes. A superior pectin derived from the seeds is capable of forming gels over a wide pH range. It has been used commercially as a creaming or jelling agent in fruit preserves, ice cream, mayonnaise and pharmaceuticals. The residual seedcake is useful as livestock feed.

Toxic The seed husk has been used at a concentration of 0.04% to eliminate undesirable species of fish from Indian ponds. The soluble saponins in the husk are believed to be the toxic property giving a response similar to rotenone, thus leaving the affected fishes edible and the aquatic insects and molluscs unaffected.

Physical The attractive hard dark purplish-brown wood is too heavy to float and has been used to make furniture, axles, mallets, tool handles, mortars and pestles. When used as fuel it produces much heat. It has also been used as a source of high-quality charcoal suitable for use as an ingredient in gunpowder. A red dye has been produced from the leaves which turns indigo-dyed cloth yellow. The pods, bark and galls are also used to produce leather and fabric dyes. The fruit pulp is used as a fixative in dyeing. The seed kernel yields a pectin and a mucilage, each with many industrial uses such as sizing in textiles. The extractives of the seed testa are used as a component of the paraformalehyde glue used in the manufacture of plywood. The seed also yields an amber oil used in food preparation and as a base for paints and varnish.

Ecological In India the tamarind is planted as a firebreak because the dense shade usually leaves bare ground under the tree. A row of these trees also makes an effective windbreak.

Notes Tamarind trees may live to be over 300 years old and their presence in the forest is often a clue to the location of 18th- and 19th-century house sites. The flowers are attractive to bees and produce a quality honey. Annual fruit yields of 200 kg (440 lb) per tree are average but yields of up to 500 kg (1100 lb) have been recorded. If the fruit is left on the tree for too long after ripening it may be attacked by weevils and other vermin. The ripe fruit contains the following constituents per 100 g (4 oz): 239 calories, 3 g protein, 0.6 g fat, 62 g carbohydrate (about 70% glucose and 30% fructose), 5.1 g fiber, 10 to 15 g tartaric acid, 8 g potassium tartrate, 74 mg calcium, 113 mg phosphorus, 2.8 mg iron, 51 mg sodium, 781 mg potassium, 86 mg ascorbic acid (vitamin C), as well as citric, malic, acetic and nicotinic acids. The fruit also contains the free amino acids proline, serine, alanine, phenylalanine and leucine. The most abundant volatile component is 2-acetyl-furan which, combined with over 60 other compounds, contributes to its citrus (limonene, terpinen-4-ol, neral), spicy (methyl salicylate, safrole, ionones, cinnamaldehyde) and roasted (pyrazines and lower alkylthiazoles) flavors. The leaves are rich in tartaric, alpha-oxo-glutaric, glyoxilic, oxalo-acetic and oxalosuccinic acids. The tartaric acid moves to the sap from the leaves then decreases in concentration as it accumulates in the ripening fruit. Unlike the acid in many fruits, tartaric acid is not utilized as the fruit matures. The leaves also contain the flavanoid glycocides vitexin, isovitexin, orientin and isoorientin. The seed makes up about 35% of the weight of fruit and is composed of 30% seed coat and 70% endosperm kernel. The seed kernel contains 65% nonfiber carbohydrates, 5.6% crude fiber, 7% fat, 17% protein and 2.8% ash. The seed oil is a mixture of glycerides of saturated and unsaturated fatty acids including oleic and linoleic. The mucilages and pectins in the seeds are composed of glucouronic and galactouronic acids, glucose, galactose, arabinose and xylose.

COIN VINE

Dalbergia ecastaphyllum   *Fabaceae*

Geographic Distribution The genus contains about 200 species of vines, shrubs and trees. This widely dispersed species is found throughout the tropical Americas and Africa.

Form A trailing or climbing shrub or vine with stems to 8 m (25 ft) in length, sometimes forming dense thickets to 3 m (9 ft) tall. The glossy leathery leaves are alternate, 5-15 cm (2-6 in) long with short stout petioles.

Flower & Fruit Date The fragrant white to pink flowers are borne in clusters in the axils mostly in the spring and summer.

Reproduction The discoid 2-cm (1-in) diameter pod, coppery at maturity, contains a single seed that is widely dispersed by the tides.

Propagation Germination of the large seeds in moist soil is the best method of starting new plants. Transplants from the wild also produce vigorous new plants. Growth rate is fast.

Habitat & Ecological Distribution Found growing above the high-water mark on beaches and dunes, but more typically on acid salty soils among and on the upland side of mangroves. They have root nodules that fix atmospheric nitrogen.

Uses

Ornamental May be used as a border plant on saline soils or to add variety on the edge of mangrove stands.

Medicinal The leaves have been used in various medicinal teas, and the tea from the root has been used to treat sore throats. In modern medicine an extracted isoflavinoid, biochanin, has been found to function similarly to estrogen hormones, perhaps explaining the folk use of the plant to induce menstruation. A group of experimental antibiotics called dalbergions has been extracted from several plants of this genus.

Toxic Like the related Jamaica dogwood, the crushed roots and bark release a compound into water which has an immobilizing effect on fish.

Physical The bark separates into long strong fibrous strips which may be used as rope for lashing timbers or branches together in primitive construction. The wood has a pleasant rose-like smell and in addition to being carved into various small items such as cutlery and brush handles, it has been used as a component of incense.

Notes The genus was named in honor of Nils and Carl Dalberg, botanists who were associates of Linnaeus in the late 18th century. The species *ecastaphyllum* refers to the unifoliate leaves. This genus also contains the rosewood tree known for its beautiful lumber. Several compounds of the 4-phenylcoumarin group have been isolated from the wood and may account for its durable qualities. Also present in the wood are prunetin, chalcones, pterocarpans, furan, benzophenone, styrene, sterol, neoflavonoids, isoflavones, isoflavans and the isoflavinoids formononetin, mucronulatol and dimethylduartin.

BEACH PEA

Canavalia rosea   Fabaceae

Geographic Distribution The genus has 51 species in the tropics and subtropics worldwide. This species is found on tropical shores and sometimes inland worldwide.

Form A prostrate, trailing, twining vine with a densely spreading root system and alternate leaves composed of 3 leaflets. The leaflets lie open and flat on the ground early in the morning, but as the sun rises higher in the sky the leaves fold along the midrib, reducing their heat-absorbing area. The vine may reach 15 m (50 ft) or more from its origin.

Flower & Fruit Date The pink to purple pea-like flowers occur in every month.

Reproduction The thick brown marbled beans are borne in robust pods to 15 cm (6 in) in length and germinate as the primary source of reproduction. The rooting of stems is an effective source of spreading and vegetative reproduction.

Propagation If the seeds are picked when mature but still green and planted immediately, the germination is quite rapid. If mature dried seeds are used they should be soaked in fresh water several days or scarified with boiling water or abrasives before planting. Germination after a week yields a vine up to 1.5 m (5 ft) long within the first month.

Habitat & Ecological Distribution The beach pea is often one of the earliest colonizers of newly deposited coastal sand and storm-battered rocky shorelines. Although it thrives in full sun, it still does very well in light shade. If a seed source is available, it will rapidly cover disturbed areas inland.

Uses

Ornamental With bright colorful flowers, a rapid growth rate and

considerable salt and drought tolerance it has considerable potential as a landscape plant. It is most used as a seaside ground cover, but it is also a good candidate for a trellis. When planted in less hostile environments than the beach, it may tend to overwhelm adjacent plants.

Edible The dry mature beans are roasted and used to adulterate coffee. The vines are consumed by livestock. The beans have a protein content of over 22% and would be suitable as a food source if supplemented with the sulfur-containing amino acids as is commonly required of other legumes.

Toxic The immature beans are considered

to be toxic and the mature beans are variously reported in the literature as edible or toxic, although investigation has not been able to find any alkaloids or cyanogenic glycosides. The dried pod has been smoked to produce hallucinations, the source of which is the compound 1-betonicine. *Ecological* The fast growth rate combined with rooting at the leaf nodes leads to sand binding for rapid control of erosion. The nitrogen-fixingnodules on the roots significantly improve the fertility of the soil. It is frequently found mixed with *Ipomoea* and seems dominant to it on the landward side of the beach.

Notes The genus is derived from a Pacific island native name for the plant. The species is named for the rose color of its flower. This vine has many synonyms, the most common of which are *C. maritima* and *C. obtusifolia*. It is also commonly called bay bean. The seeds contain the enzyme urease, which hydrolyses urea to ammonium carbonate and is used as a laboratory reagent to test the urea content of blood and urine. The seed also yields *Concanavaline A*, which clumps red blood cells and agglutinates fibroblast tumor cells, thus showing potential for future cancer research. The seeds are known to contain variable amounts of canavanine which is an analog of the amino acid L-arginine and interferes with arginine utilization. When this analog is incorporated into insect proteins it significantly disrupts numerous metabolic reactions and thus reduces future generations of seed eaters. A bruchid beetle has evolved the ability to discriminate between the analog and the amino acid and avoids synthesizing proteins with canavine. This beetle's larvae are the major predators of the seeds of bay bean. Fossils of this plant have been found in Eocene and Miocene rocks.

CORALBEAN

Erythrina corallodendrum   *Fabaceae*

Geographic Distribution The genus of about 100 species is found throughout the tropics usually as thorny shrubs or trees. The natural range of this species is from Jamaica to Trinidad, but it has been widely planted in southern Florida and Central America. The related *E. herbacea* is found in the southeast U.S.

Form A shrub or small thorny tree to 8 m (25 ft) tall with a trunk of 20 cm (8 in). The alternate compound leaves have 3 wedge-shaped leaflets.

Flower & Fruit Date The brilliant red flowers grow in a developing cone on a vertical spike in late winter or spring while the tree is still leafless.

Reproduction The black pods split open in the fall to reveal the hard glossy scarlet seeds.

Propagation The coralbean is best grown from seeds freshly picked from the tree and scarified with sandpaper before planting. Set in flats with moist potting mix, the seeds will germinate and show themselves in less than 2 weeks. Germination percentage declines with dried seeds. Cuttings  of large sticks treated with rooting hormone and set in a mixture of humus, sphagnum and sand will produce new plants most rapidly. The growth rate is moderate but improved with fertilizer.

Habitat & Ecological Distribution It naturally occurs along the coast and in low elevation dry forests but it has been planted and does well in a broad variety of habitats. It prefers full sun but does well in open shade. It is tolerant of salt and drought but not cold.

Uses

Ornamental This tree has been planted singly and in groups for the showy flowers. Its salt tolerance makes it ideal for lots composed of dredge fill.

Medicinal Various plant parts have been used (dangerously) in home remedies to treat asthma, neuralgia and other ailments. A tea prepared from the foliage has been used as a diuretic and laxative. The sleep-inducing property of the leaves has led to their use in treating hysteria. While many medically useful compounds seem to be present in this plant, a simple tea extracted from plant parts will likely also contain harmful compounds. Many people have been poisoned by the consumption of home remedies prepared from this tree. One of the alkaloids shows medical promise due to its curare-like action when injected. It has been used as a muscle relaxant in surgery and for control of convulsions in shock therapy. Another alkaloid, erythrodine, produces muscle relaxation and hypnotic effects when taken orally.

Edible Young flowers and leaves have been reported as being edible when cooked.

Toxic The leaves, bark, wood, seeds and pods have been found to contain over 30 toxic alkaloids. The seeds of many species in this genus contain an isoquinaline which induces hallucinations. The crushed seeds, stems and leaves have been used as a poison to catch fish. The crushed seeds have been used as a rat poison.

Physical The light fibrous wood is strong but not durable. The wood has been used as a substitute for cork and a media for carving. The seeds have been used in necklaces, rosaries, good luck charms and other novelties. *Ecological* It is an ideal member of a dune-stabilizing shrub community. The flowers have the classic shape and color to attract hummingbirds. It is also favored as a support for the vines of the vanilla orchid.

Notes The genus name is derived from the Greek *erythros* meaning red. The species also refers to the red flowers, comparing them to the red coral found in the Mediterranean. *Erythrina herbacea* has also been called *E. arborea*. Coraltree and cardinal spear are other commonly used English names.

JAMAICA DOGWOOD
Piscidia piscipula   *Fabaceae*

Geographic Distribution The genus of 8 species of shrubs and trees is found in the American tropics. The dogwood is found on the tropical Atlantic and Pacific coasts of the Americas.

Form A modest-sized tree with an open erect crown to 15 m (50 ft) tall and a trunk of 0.6 m (2 ft). The pinnate leaves usually have seven leaflets each 5 to 10 cm (2 to 4 in) long.

Flower & Fruit Date The pink to lavender pea-shaped flowers are borne in clusters on slender stalks in the spring. The flowers are rendered more striking by appearing between the loss of old leaves and the growth of new ones.

Reproduction The pods are 7 cm (3 in) long, contain 3 to 7 seeds, and have four papery wings. They mature in early summer and remain on the tree for many months.

Propagation The 6-mm (1/4-in) brown seeds will germinate in 8 to 10 days if planted one seed width deep in moist soil as soon as they are removed from the ripe dry pod. Cuttings root readily to the extent that green wood used as posts frequently takes root. The growth rate is rapid and is accelerated by fertilizer. Young trees may need some supplemental water to see them through their first year or so of drought. They may bloom as early as their 4th year at a height of 4 m (12 ft).

Habitat & Ecological Distribution The dogwood is found in a variety of dry low-elevation habitats

beginning in the sand at the upper edge of the wave wash zone and continuing inland on the dry clay of Caribbean hillsides and the rich moist humus of South Florida hammocks. It is salt- and drought-tolerant but sensitive to cold.

Uses

Ornamental The dogwood is grown in yards and along fence lines as ahardy shade tree with colorful flowers.

Medicinal The bark or crushed leaves have been used to relieve pain. The dried root bark has been used to treat neuralgia, insomnia, hysteria, toothache, asthma, and alcoholism. An extract of the bark has been used to treat mange in dogs. Research has shown that the primary active ingredient in the bark is piscidic acid ($C_{11}H_{12}O_7$) and that it has narcotic

effects, causesdilation of the pupils, increased perspiration and a rise in blood pressure. Its actions have been found similar to both morphine and cannabis. Other researchers have found that it lowers the amplitude of uterine and intestinal contractions. Generally, extracts have been found to have antitussive, antipyretic and antispasmodic action. Extracts are used in several proprietary female tonics.

Toxic Indians have used the bruised root bark, twigs and leaves to temporarily stun fish. The fruits have been used as a component of arrow poison. Extracts and powders of the plant have also been found to be potent insecticides.

Physical The yellow-brown wood is hard, heavy, tough and resistant to decay. It is difficult to work but finished articles are attractive and take a high polish. It has been used for making furniture, carts, wheel spokes, boat parts, posts and makes excellent charcoal.

Ecological The seasonally abundant flowers have been shown to be a major source of pollen (and presumably nectar) for honeybees.

Notes The generic name is derived from Latin words meaning "fish killer." The species name translated from Latin means "little fish." This plant has also been called *Piscidia erythrina* and *Ichthyomethia piscipula.* The tree is also known by the common name fishpoison tree. The compounds in the root which are toxic to fish are many and varied. Ichthynone with the formula ($C_{23}H_{20}O_7$) is lethal to fish at one part per million. Rotenone, the well known fish killer used by native peoples even prior to European invasion, is present along with the similar compounds sumatrol, millettone, isomillettone, and dehydromillettone. Also present is piscidic acid and its esters, the isoflavonoid lisetin, two isoflavones (piscerythrone and piscidone), fukic acid, beta-sitosterol, malic, succinic, and tartaric acids, a saponin glycoside, tannin, jamaicin, waxes and others. Various combinations of these chemicals are also present in the stems and leaves.

Lignumvitae

Guaiacum officinale　　　*Zygophyllaceae*

Geographic Distribution *Guaiacum officinale* is found in the West Indies and northern South America. The related *G. sanctum* occurs in Florida, Central America and the Greater Antilles.

Form A small tree to 9 m (30 ft) tall with a deep-green, dense, rounded crown and a trunk up to 70 cm (30 in) but usually less than 30 cm (12 in) in diameter. The bark is thin and smooth with irregular glossy flakes. The bark of *G. sanctum* is rough. The pinnate leaves have 4 to 6 oval leaflets without petioles. *G. sanctum* has 6 to 10 paired leaflets. Very old gnarled trees have been determined to be over 1000 years old.

Flower & Fruit Date Both species have small fragrant flowers with five blue petals.

Reproduction Both species have seed capsules, orange when ripe, splitting open to reveal a bright red covering over dark brown seeds. G. *sanctum* has 4 or 5 angled or winged fruit, each section containing a seed. *G. officinale* has a flattened heart-shaped capsule containing two seeds.

Propagation Lignumvitae seeds sprout eagerly but grow very slowly. It is not uncommon to have 100-year-old trees with trunks only 12 cm (5 in) in diameter. Gnarled individual trees over 1000 years old have been

recorded. Recently it has been discovered that weekly spraying of the foliage with the growth hormone gibberellic acid at a concentration of 0.01% accelerates the growth by a factor of 4. While this treatment would not be practical for established trees it might help newly set out plants reach a noticeable size in a shorter time period.

Habitat & Ecological Distribution Lignumvitae is found from the seashore inland in dry coastal forests. It manages to thrive and maintain a healthy appearance when exposed to drought, salt, and a variety of discouraging soil conditions. This is one of the climax species of the natural dry coastal forest as it was found by the first Europeans in the New World.

Uses

Ornamental These trees are planted as ornamentals for their hardiness, deep green foliage, attractive blue flowers and cascades of orange fruits.
Medicinal The New World explorers carried syphilis back to Europe from the West Indies in their early voyages. As a result of observing the Indians on Hispaniola treating this disease with lignumvitae, a great demand for this wood in Europe developed. Beginning in 1508 and for over 2 centuries the wood was used to treat the syphilitic sores and many other diseases. The saponins in the resin probably did help reduce the external lesions and somewhat inhibit the causative spirochetes. More recently home remedies have used the leaf tea to treat rheumatism, asthma, high blood pressure and diabetes. The bark, wood and resin are used in folk medicine to treat gonorrhea, gout, rheumatism, and to induce menstrual flow or abortions. The resin is known in commerce as guaiac and is used in Eastern medicine to treat rheumatoid arthritis and several mucous membrane diseases.

Edible The fruit has been reported as being edible when cooked as a vegetable. The resinous extract of the wood is 15% vanillin and is used as a flavoring in nonalcoholic beverages, ice cream, candy, baked goods, gelatins, puddings and chewing gum.
Physical The wood is much harder than other American woods. It is heavy with a fine and uniform texture and an interlocking grain. It is generally difficult to work but takes a high polish. The specific gravity of 1.3 is so high that it sinks, even in seawater. The resin content of the wood (up to 25%) and resistance to wear have caused it to be used as self-lubricating bearings for the propeller shafts of ships and water-driven turbines. It is used in industrial bearings when contact with acids, alkalies or water would destroy most metal bearings. Over 100,000 railroad crossties of lignumvitae were used in the construction of the Panama railroad in the late 19th century. It has also been used for pulley sheaves, rollers, casters, mallets and bowling balls. It is highly resistant to marine borers. The resinous oil extract with a woody roselike note is used in perfumery.
Ecological This is one of the most hurricane-resistant trees available for landscape planting. The arils of the fruit are consumed by many species of birds which also disperse the seeds through the forest.
Notes The generic name is based on the original Indian name *guayacan*. The species *sanctum* is derived from the mistaken recorded occurrence of the

tree on the island of Saint John. Bees are enthusiastic about the flowers. The tree is also known as guayacan and ironwood. Medicinal demand for the wood in the 16th and 17th centuries led to such extensive logging and export of this tree that most large stands were extirpated, leaving only a few trees scattered in the forest and almost eliminating the tree in many parts of its range. The resin is extracted from sawdust by boiling it in salt water or by boring a hole in the log, heating the wood and collecting the resin as it runs out. The resin is marketed as gum guaiac or guaicum resin and is listed as a drug in the British Pharmacopoeia. It is generally recognized as safe by the U.S. Food and Drug Administration as a preservative and antioxidant in fats at a concentration of up to 0.10%. The resin contains 70% guaiaconic acid, 10% guaiaretic acid, and numerous other phenolic compounds belonging to the lignan group. Tincture of guaiac turns blue in the presence of many oxidizing agents and has been used as a test to detect oxidative enzymes, blood stains and occult blood. It is also used in forensics to distinguish the blood of birds, fish and reptiles from that of mammals. The blue color has been identified as the oxidation product of a-guaiaconic acid. The chemistry of guaiac, guaiacol and related compounds is more extensively defined in the Merck Index. The fruit and stem bark have a series of unusual nortriterpenoid saponins, larreagenin, oleanic acid and sitosterol.

LIMEBERRY
Triphasia trifolia   *Rutaceae*

Geographic Distribution The single species in the genus is a native of Southeast Asia and has been cultivated and naturalized in much of the tropics. It is well established on the Gulf coast and throughout the Caribbean.

Form A thorny evergreen shrub with a variable growth habit sometimes forming a small tree to 5 m (15 ft). The shiny leaves are composed of 3 leaflets with wavy edges and very short petioles. The central leaflet is about twice the size of the other two. At the base of each leaf is a pair of short slender straight sharp spines. The foliage is aromatic with a fragrance typical of citrus when crushed.

Flower & Fruit Date The small fragrant white flowers are borne at the base of leaves throughout the year.

Reproduction Upon ripening, the 12-mm (1/2-in) diameter thick-skinned berries turn dark crimson with a mucilaginous spicy pulp enclosing a single seed.

Propagation Seeds and cuttings have both been used successfully in starting new plants.

Habitat & Ecological Distribution A hardy plant which can be found thriving on soils from carbonate sand to heavy clay and gravel. It can tolerate full sun but seems to do best under the open shade of tall trees. It is quite frost sensitive. Its dense vigorous growth under the canopy of other mature trees can cause it to be a nuisance which crowds out native vegetation.

Uses

Ornamental It is widely planted as an ornamental shrub. The compact dense foliage lends itself well to use as a trimmed hedge or a low-maintenance privacy screen. If left undisturbed it forms dense impenetrable thorny thickets.

Medicinal A medicine to treat diseases of the chest is prepared from the fruit. The leaves have been used to treat diarrhea, colic and skin diseases.

Edible The fruits are pleasant when eaten raw and can be made into marmalade or jam, preserved whole in a sugar syrup or used as an interesting filling in sweet pastries. An aromatic, spicy flavored liqueur may be made by soaking the intact fruits in strong spirits for several years.

Physical The leaves have been used as a component in cosmetics, and the fruit gum has been used as a glue. The wood is very hard and resilient but due to small size is seldom used except for tool handles and small rustic furniture parts.

Ecological The seeds are widely distributed by birds which eagerly consume the ripe fruit.

Notes *Triphasia* refers to the 3-part flowers and *trifolia* refers to the 3 leaflets per leaf. This plant is also known as sweetlime, Chinese lemon and chinita. The flowers are known to be a good source of honey. The mature fruit contains large amounts of beta-carotene and derivatives which decline in concentration as the fruit ripens. These are replaced as the principal pigment by the rare carotenoid triphasiaxanthin found only in this species and in cycads. Analysis of the leaves shows the presence of numerous alkaloids, steroids, terpenoids and saponins.

LEATHERLEAF

Stigmaphyllon periplocifolium   *Malpighiaceae*

Geographic Distribution The genus has about 100 species found in the warm parts of the Americas. This species is found throughout the Caribbean.

Form A tough woody vine reaching a length of 15 m (50 ft) and a thickness of over 3 cm (1 in). The elongate oval leathery leaves are opposite and typically about 8 cm (3 in) long.

Flower & Fruit Date The bright yellow flowers are borne in clusters mostly in the spring, but also in response to rainfall.

Reproduction The reddish, winged, paired seeds mature in the summer and fall.

Propagation This plant germinates and grows well from seed collected from wild plants.

Habitat & Ecological Distribution Often one of the few plants thriving on small bare rocky islands. It may also be found in almost every other dry habitat except bare sand.

Uses

Ornamental The hardy long-lived nature of this plant suits it for covering fences, arbors, or rocky areas with a carpet of green which is very resistant to drought and yields periodic outbursts of flowers.

Medicinal An extract from the leaves and stems has been used as a shampoo for dandruff. Tea from the bark has been used to treat strain, back pain and a run-down system.

Ecological It provides cover and nest sites for many species of animals on otherwise barren habitat.

Notes The genus is from the Greek and is named for the leafy stigma. The species is named for the rounded leaves with curved margins. *Stigmaphyllon emarginatum* is another accepted name for this plant. Soldier vine and snake root are other common names. The iridoids galioside, monotropein and geniposidic acid have been isolated from members of this genus.

BAY CEDAR
Suriana maritima   *Surianaceae*

Geographic Distribution This is the only species in the family and it occurs on tropical shores worldwide.

Form A small shrub with narrow succulent leaves, sturdy branching trunks and slim erect branches forming rounded clumps. Sometimes forming a small tree to 6 m (20 ft) in height with a trunk to 12 cm (5 in) in diameter. The grayish succulent leaves are bunched near the ends of twigs. The twigs and leaves are covered with a fine fuzz.

Flower & Fruit Date The small yellow 5-petaled flowers with conspicuous green sepals may occur in any month of the year. One branch of a plant may be setting fruit while another branch still has flowers in bud.

Reproduction The four or five fuzzy floating fruits arising from each flower are conical and contain a black finely haired seed.

Propagation Small seedlings can be transplanted with good results if the roots are kept wet until replanted. Care should be taken to keep the long taproot straight when planting and it should not be given a post-transplant pruning. Sprinkle seeds on the surface of a light potting mix or pure sand and expect germination in about 2 or 3 weeks, but be patient as it sometimes takes much longer. A modest salt content in the media (or a commercial fungicide) helps prevent the damping-off fungus. Seed viability is a problem with many source plants.

Habitat & Ecological Distribution

The bay cedar grows on sandy shores in full sun, often directly fronting the wave wash zones.

Uses

Ornamental The extreme salt tolerance and low growth habit make this plant an ideal choice for a low hedge on the beach front. Its great potential as a wind-resistant landscape plant has seldom been recognized, but a few native plant nurseries are now making it available.

Medicinal Several different methods of preparing the foliage have been used in home remedies to treat intestinal problems. A tea from the leaves and bark has been used to clean sores and treat fevers. The flowers are made into a tea to treat blood disorders.

Physical The reddish-brown wood is hard, heavy and of a fine uniform texture. It has been carved into spears and fish hooks but is seldom large enough to be useful for lumber. The reddish-brown resin exuding from cut surfaces has been used as a sealant.

Ecological This plant is a valuable component of mixed species dune-stabilizing plantings.

Notes The genus is named for Joseph Surian, an 18th-century French botanist.

The species name *maritima* refers to its growing near the sea. The wood contains beta-sitosterol, various alkane hydrocarbons, and a lignan composed of vanillin, syringaldehyde and several aromatic acids.

GUMBO-LIMBO
Bursera simaruba   *Burseraceae*

Geographic Distribution The gumbo-limbo is found in South Florida, throughout the Caribbean and on the Pacific coast of Central America.

Form A medium-sized tree sometimes growing to 27 m (90 ft) and 1 m (3 ft) in diameter on ideal sites. It has a shiny reddish bark which scales off in papery flakes (thus called tourist tree "because it is always red and peeling"). The red sheen of the bark becomes more silvery under good growing conditions. The pinnate leaves have 3 to 7 opposite leaflets. The tree forms deep tap roots if the substrate allows and does not generally damage nearby structures.

Flower & Fruit Date The small inconspicuous white flowers are borne in clusters mostly in the spring. While the flowers last only one day, they begin nectar production before dawn, leading to intense honeybee activity at daybreak.

Reproduction The bunches of thin-fleshed, dark-red fruits with large whitish seeds may grow to full size in a week but take an additional 8 months before they mature and ripen in the summer. Trees begin to produce seed when 5 years old.

Propagation The cut branches up to 10 cm (4 in) in diameter placed in suitable soil root readily. Plants grown from seed or transplanted seedlings have a faster growth rate and form more symmetrical trees. A broader more bushy tree can be produced if the vigorous terminal sprout is pruned. The growth rate is faster with light fertilizer. There is a large variation in viability of seeds from different trees.

Habitat & Ecological Distribution The gumbo-limbo is very tolerant of salt, drought and poor soil. It is remarkably fast growing in the face of adversity. The trunks shrink during the dry season and expand again as they store water in the rainy season. Mature trees can stand occasional light

frosts. In spite of their notoriously weak wood, their sturdy growth form makes them very resistant to structural damage from hurricanes.

Uses

Ornamental The attractive red bark and moderately dense canopy have made this a fully accepted native street tree or shade tree. Nurseries are increasingly making this tree available for home landscaping.

Medicinal The resin has been used medicinally and as a plaster to cover contusions and other injuries. Teas prepared from the leaves and bark are astringent due to a high tannin content and are used in a great variety of home remedies including the treatment of rashes, diarrhea, bruises, back pain, strain, tiredness and fever. Combined with lignum vitae it is a treatment for "weakness in men."

Edible The leaves have been used to produce a refreshing aromatic tea and the resin has been used to flavor confections. The young tender shoots may be cooked as a vegetable.

Physical The lightweight, tough, fine-textured wood has been used for carving, matchboxes, toothpicks, carousel horses, and is a traditional material for voodoo drums. It has ideal physical and mechanical properties for use as matches. It is used as a veneer for plywood and sometimes has a birdseye figure similar to maple. The wood is very subject to termite attack unless treated. Pulp produced from the wood is suitable for production of printing and writing paper. Cut branches used as fence posts often take root, producing a living post. Lines of these trees in the countryside have resulted from this practice. The resin produced by injuring the tree bark (called *copal, chibou, cachibou, gomart* or *gum elemi*) was highly valued by the pre-Columbian Mayas medicinally, as incense, and as a varnish coating for canoes. The resin has also been used as an insect repellent and as an adhesive to repair crockery. Today it is used in lacquers and varnishes, and as an incense component.

Ecological The fruits are consumed by birds, rodents and other wildlife. It is thought that the hard seeds serve the same function as pebbles in a bird's gizzard. Trees on dry sites are red, while those with more water and nutrients grow faster and produce a more silvery bark.

Notes The genus is named for Joachim Burser, a 17th-century German botanist. The species *simaruba* is from an Indian name for the tree. The common name gumbo-limbo is derived from the African Bantu language. In the West Indies this is often called turpentine tree due to the resinous smell produced by several volatile terpenes in the bark. Frankincense and myrrh are produced from related trees in Africa. This tree continues to photosynthesize in the absence of leaves by using chloroplasts under the bark surface. The wood is marketed under the name West Indian birch and the plywood has been marketed as Mexican white birch.

BUSHY SPURGE
Euphorbia articulata   *Euphorbiaceae*

Geographic Distribution   The genus has hundreds of species found worldwide. This species is found from the Bahamas through Puerto Rico and the Lesser Antilles.

Form   A small shrub, sometimes sprawling and reaching a length of 6 m (20 ft). The oblong opposite leaves may reach a length of 6 cm (2 1/2 in) long by 1 cm (3/8 in) wide, but are usually about half that size.

Flower & Fruit   The small inconspicuous flowers are borne in a cuplike structure called an involucre. The three angled seed capsule is fuzzy and about 3 mm in length.

Reproduction   The seeds are grayish pink with transverse ridges.

Propagation   This plant may be propagated by cuttings.

Habitat & Ecological Distribution   It may be found on rough rocky shorelines on cobble beaches and inland among scrubby vegetation.

Uses

Ornamental   Although they are seldom selected by landscape architects, these hardy plants are an ideal candidate for providing a bit of green vegetation on harsh rocky shorelines.

Medicinal   Many *Euphorbia* have been used to treat cancers, tumors and warts for over 2000 years. Modern medicine has found an extract which shows significant inhibitory action against several sarcomas, carcinomas and leukemia in mice.

Toxic   The sap is a caustic irritant, often causing dermatitis and blistering of the skin. A drop of the sap in the eye causes burning and swelling of the eyelids, sometimes with temporary blindness due to corneal clouding. Internally, the *Euphorbia* cause burning and blistering of the lips, tongue and throat, vomiting, abdominal pain and urinary irritation. Nervousness and hysteria are often side effects of ingestion. Coma and even death are recorded from ingesting the latex.

Notes   The genus is named for Euphorbus, a Greek physician. The species *articulata* means jointed. The name *Chamaesyce* is also used for this genus of plants. This plant is also called seaside spurge. All the *Euphorbia* have toxic, irritating white sap containing the resinous triterpene euphol, ($C_{30}H_{48}O$), euphorbin ($C_{29}H_{36}O_7$), euphorbol or tirucallol. The compound active against tumors is a diterpinoid diester, ingenol 3.20 dibenzoate. A compound active against leukemia, phorbol-12 tiglate-13 decanoate, has been identified.

COAST SPURGE
Euphorbia mesembrianthemifolia   *Euphorbiaceae*

Geographic Distribution The genus, which contains hundreds of species, is found worldwide. This species is found from Florida through Central America and the West Indies.

Form A bushy, upright shrub or herb to 60 cm (2 ft) tall. The oval, opposite 9-mm (3/8-in) leaves and their supporting stems yield a thick white latex.

Flower & Fruit Date The minute white flowers may be found in most months on the branch tips.

Reproduction The tiny white 3 or 4 angled seeds are born in a capsule.

Propagation This hardy little plant will root from cuttings or germinate and grow from seed.

Habitat & Ecological Distribution The coast spurge is usually found on dry sites on both sandy and rocky coastlines.

Uses

Ornamental This is a good candidate to add some variety to low coastal plantings.

Medicinal The latex has been applied to sea urchin punctures and other wounds. A tea prepared from the foliage has been used in home remedies. The seeds are both extremely purgative and emetic and an overdose may cause nervousness, coma or death. Historically the "Croton oil" from the seeds was sold and used as a purgative, but it is now generally abandoned as being too violent and a co-carcinogen.

Toxic All parts of the fresh or dried plant are toxic. The plant is reported to have been lethal to horses and other livestock that graze on it. Grazing animals consuming small amounts develop swelling around the eyes and mouth, burning in the mouth and throat and abdominal pain. Cows give a bitter reddish milk after consuming small amounts. Goats can eat

Euphorbia but the milk they give causes diarrhea in humans. Livestock have been poisoned by water containing *Euphorbia* leaves.

Ecological This is a useful sand binder to add ecological diversity to sandy shore communities.

Notes The species name refers to the leaves resembling those of the plant *mesembrianthemum*. This plant is also known as *Chamaesyce buxifolia*. The honey produced from *Euphorbia* flowers is bitter and acrid, causing burning in the mouth and throat, vomiting and diarrhea. Croton oil from the leaves, stems and seeds is composed primarily of the benign oleic, linleic, palmitic and stearic fatty acids. The purgative irritant is 12-0-tetradecanoyl-phorbol-13-acetate.

MANCHINEEL

Hippomane mancinella   *Euphorbiaceae*

Geographic Distribution The single species in the genus occurs throughout the American tropics and has been introduced into west Africa.

Form A medium-sized tree with spreading branches and a rounded crown to 20 m (65 ft) in height with a 1-m (3-ft) diameter trunk. The oval leaves are opposite with glands on the petioles and oval or lightly toothed margins.

Flower & Fruit Date The small greenish flowers are borne on a spike in the summer.

Reproduction The small applelike fruits, yellow when ripe, are borne in the forks of twigs. The fruit has a pleasant aroma and a sweet taste. The seeds are viable and dispersed after passing through the gut of various birds and herbivorous lizards.

Propagation Although the seeds sprout and grow well it is recommended that they be destroyed. This tree should not be intentionally propagated in human use areas.

Habitat & Ecological Distribution The manchineel thrives in dry, sandy, salty coastal communities and often on berms and higher land inland of mangroves.

Uses

Ornamental While the tree would make an attractive and salt-tolerant beachfront source of shade, it is probably too hazardous for most locations.

Medicinal Extracts of the seeds and bark have been used as a vermifuge,

85

as a cathartic, and to treat venereal diseases and tetanus. The treatment was probably worse than the disease.

Edible In spite of the extremely adverse aftereffects, the fruit is sweet, juicy and pleasantly aromatic to eat.

Toxic All parts of the plant are toxic. The milky sap is highly irritating, causing a rash and blistering on the skin and temporary blindness in the eye. If the fruit is chewed and swallowed, the effects may not be evident for several hours. Pain, salivation and swelling of the lips will be accompanied by a loss of the surface tissue of the tongue, gums and throat. Abdominal pain, vomiting and bleeding of the digestive tract are usual. Shock and death may occur without supportive therapy. Chronic or small doses may be lethal over time due to hemorrhage and degeneration of the liver, kidneys, pancreas and adrenals. The aerosol produced by chopping the tree is an irritant and the smoke from burning leaves or wood is irritating to the skin, eyes and respiratory system and induces a severe headache. The various toxic properties have been lethal to both humans and livestock. One of the toxic components is water soluble, thus abundant rinsing of the exposed area helps to reduce the external effect. As treatment for the ingestion of manchineel, lavage with oil followed by a saline cathartic has been recommended. The fall in blood pressure can be antidoted with epinephrine or ephedrine.

Physical Although its toxic nature presents difficulties, the strong dark-brown wood with yellow variegations resembling Circassian walnut takes a good polish and has been used for furniture, cabinets, and construction.

Ecological The capability of growing close to the sea on a variety of soil types makes this tree a useful plant to stabilize shorelines in areas infrequently used by humans. Birds and land crabs feed on the fruit with seeming impunity, but the latter are said to be rendered toxic for humans after eating manchineel.

Notes The genus name is from the Greek meaning "horse poison." The species name means "little apple." Columbus noted that the Carib Indians used the milky sap to poison arrows. Horatio Nelson was poisoned by manchineel when hostile Indians put leaves in a spring which he used as a water source. In most developed areas this tree has been eliminated due to its potential for producing health problems. The toxins are quite varied; some are water soluble and others soluble only in organic solvents; some produce an instantaneous irritation while others take some time to make their influence felt, and still others induce the future formation of benign and malignant tumors. Many of the toxic compounds have not been chemically identified, but the water-soluble toxic principle of the fruit seems to be the alkaloid physostigmine. A series of compounds in the leaves are methyl derivatives of ellagic acid. Hippomanin A, a water-soluble highly toxic compound from an ethanol extract, is a pale yellow crystalline solid with the formula $C_{27}H_{22}O_{18}$. Bees avidly seek the abundant nectar which produces a surprisingly good (nontoxic) honey.

CROTON

Croton discolor   *Euphorbiaceae*

Geographic Distribution The genus has 600 species of plants from herbs to trees that occur worldwide. This species is found in the Bahamas and Caribbean.

Form An erect perennial shrub to 2 m (6 ft) tall. The oblong alternate leaves are dark green and smooth on top, light green and hairy below. The flowers, twigs and leaf stems all have starlike white hairs.

Flower & Fruit Date The small inconspicuous flowers of each gender are borne on separate stems throughout the year.

Reproduction The hairy spherical fruits are about 6 mm (1/4 in) in diameter and contain 3 seeds.

Propagation The seeds sprout and grow with such vigor that this plant has become a pasture pest.

Habitat & Ecological Distribution This plant is very tolerant of salt, wind and dry conditions. It thrives best in full sun but survives well in open shade.

Uses

Ornamental This sturdy little shrub has potential as a short border plant or as a background for a bed of low succulents or cacti.

Medicinal Young leaves and branch tips have been used in a tea to treat coughs. The fresh sap has been used to cover fresh wounds. A component of croton oil (phorbol 12-tiglate 13 decanoate) has been found to inhibit leukemia.

Toxic The toxic purgative "croton oil" is found in the seeds, leaves and stems. While it is effective in small doses, more than a few drops may be fatal to humans or animals due to severe gastroenteritis. External application of the oil may cause blistering of the skin. The esters of thetetracyclic diterpenoid phorbol found in croton oil have been found to promote tumors in mouse skin previously treated with subcarcinogenic doses of carcinogens.

Ecological Because it is one of the few pasture plants that goats refuse to eat, abundant stands frequently indicate a landscape severely overgrazed by free-ranging goats.

Notes The genus is derived from the Greek name for the castor bean. The species name refers to the upper and lower surfaces of the leaf being of different colors. This plant is also called maran. Croton oil contains 37% oleic acid, 19% linoleic acid, 7% myristic acid, 2% arachidic acid and smaller amounts of stearic, palmitic, lauric, valeric, tiglic, butyric, acetic and formic acids. The toxic component is less than 4% of the total volume. The alkaloids crotonosine, 8,14-dihydrosalutaridine, L-N-methyl-crotonosine, linearisine and discolorine have been identified from the foliage.

CHRISTMAS BUSH
Comocladia dodonaea   *Anacardiaceae*

Geographic Distribution The species is found throughout the greater and lesser Antilles from Hispaniola to St. Vincent.

Form A shrub sometimes growing in a vinelike habit to 6 m (20 ft) with a trunk to 15 cm (6 in) and dark-green alternate pinnately compound leaves to 15 cm long. The stalkless toothed leaflets with three slender sharp spines, often tinged with red, make the tree reminiscent of Christmas holly.

Flower & Fruit Date The minute 2-mm (1/16-inch) dark reddish-purple flowers are borne on a short lateral stalk intermittently throughout the year.

Reproduction The orange-red fleshy elliptical 10-mm (3/8-in) fruits contain a single seed.

Propagation Growing in scattered colonies, it seems to readily self-seed.

Habitat & Ecological Distribution This hardy shrub grows well on beach berms and in dry and moist forest. It does poorly in shade.

Uses

Ornamental It is very decorative and is becoming increasingly available from nurseries.

Medicinal Teas from the leaves have been used to treat colds and fever. Externally the foliage has been used to prepare an invigorating bath. The bath might provoke a rash over the entire body of a person sensitive to the toxins.

Edible Pearly-eyed thrashers have been seen to eat the fruit, but most animals probably avoid the fruit and foliage.

Toxic Contact with the plant or its sap by sensitive persons causes itching, burning skin eruptions similar to those caused by its relative, poison ivy.

The irritation peaks several days after touching the plant; thus a person can absorb enough toxin to cause an explosion of irritation without being aware of exposure. The toxin can be carried through the body and erupt elsewhere than the point of contact. Persons not initially reacting to this plant may become sensitized with repeated exposure.

Physical The very attractive dark red-brown wood is hard and heavy but seldom reaches lumber size. It is used for carving candlesticks, napkin rings, and other small decorative items. The sap turns black on exposure to air and makes a lasting stain.

Ecological This is often a significant component of natural plant communities on steep dry rocky hillsides.

Notes This plant has many common names in English, Spanish, French and local dialects. A common Spanish name for this plant is *chicharon*. The irritating chemical is an oleoresin called urushiol which contains 3 pentadecylcatechol.

Marble tree

Cassine xylocarpa   *Celastraceae*

Geographic Distribution The genus of about 40 species is found in the tropics worldwide but primarily in the Old World. The marble tree is found throughout the Caribbean.

Form Usually a shrub, but may at times form a tree with a narrow erect crown to 10 m (35 ft) in height with a trunk to 20 cm (8 in). The thick stiff leaves are generally elliptic but vary considerably in size and shape.

Flower & Fruit Date The small greenish-white flowers may be found in clusters near branch tips intermittently throughout the year.

Reproduction The hard yellow-orange fruits from which the common name is derived are spherical to elliptic and about 1 inch in length with a very hard thick-walled stone.

Propagation With patience the sturdy nut will germinate and produce a vigorous young plant. When grown in pots the plants often need stakes to help stabilize them until their root systems develop fully.

Habitat & Ecological Distribution The marble tree may grow with its roots in seawater-soaked sand but also thrives at low elevations on all soil types in dry areas.

Uses

Ornamental The hardy

habit with attractive foliage and colorful fruits make this an ideal candidate for a border planting. The shrub continues to stand hale and hearty with no outward signs of distress in the most severe droughts.

Physical The fine-textured light-brown wood is hard, heavy, and strong.

Ecological The extensive root system has been demonstrated to have a good stabilizing effect protecting the face of coastal dunes against hurricane-induced erosion. With over half of the root structure exposed by hurricane erosion, the trees on one coastal dune continue to thrive.

Notes The genus is an Indian name for this plant. The species name refers to the woody fruit. The name *Elaeodendrum* has also been used for this group of plants. Olive wood is another common name for the plant.

Poison cherry
Crossopetalum rhacoma   *Celastraceae*

Geographic Distribution The genus of about 20 species is found in the American tropics and subtropics. This species is found in Florida, the Bahamas and the Caribbean.

Form This shrub or small tree grows to 8 m (26 ft) tall with a diameter of 8 cm (3 in). The opposite rounded oval leaves have petioles less than 2 mm (1/16 in) long.

Flower & Fruit Date The tiny greenish-red four-parted flowers less than 3 mm (1/8 in) across occur in clusters at the leaf bases in any month of the year.

Reproduction The 6-mm (1/4-in) long egg-shaped fruits are bright red to burgundy and contain a single seed.

Propagation Seeds germinate readily and grow into healthy plants in pots.

Habitat & Ecological Distribution Found from just above high-water mark on beaches to an elevation of 200 m (600 ft) in both wet and dry forests. It is moderately tolerant of open shade but does well in full sun.

Uses

Ornamental It is seldom planted as an ornamental but has great potential with its attractive petite foliage and bright red berries.

Medicinal Folklore has attributed several medicinal virtues to this plant, the most prominent of which are the stimulation of urine flow, treatment of kidney inflammation and expulsion of kidney stones.

Physical The hard light-brown wood is seldom used except as posts.
Ecological The open canopy growth habit would allow its use as a modest windbreak along a beach berm while allowing views of the ocean.

Notes The genus name means "fringed petals" and the species name was used by Pliny for another Old World plant. This shrub is also known as *Rhacoma crossopetalum.*

Seaside maho

Thespesia populnea   *Malvaceae*

Geographic Distribution The genus of over a dozen species is found primarily in Africa and Asia. This species is a native of the Old World tropics and has now been widely planted and naturalized on warm shores throughout the world.

Form A small tree with a dense crown of glossy heart-shaped leaves and spreading lower branches sometimes reaching a height of 18 m (60 ft) with a trunk 20 cm (8 in) in diameter.

Flower & Fruit Date The 5-cm (2-in) yellow bell-shaped flowers, with deep red centers, turning purple late in the day, occur laterally at leaf bases throughout the year.

Reproduction The five-parted brown fruit remains attached to the tree for some time after maturity. Both the unopened fruit and the seeds float and are widely dispersed by waves and currents, sometimes forming forests of seedlings at the high-tide line.

Propagation The seeds germinate readily in moist sand. Propagation by cuttings is also possible but damp-off is a problem unless the rooting medium is kept fairly dry or includes some salt. Air layering also works well.

Habitat & Ecological Distribution This is a vigorous salt-tolerant tree which is able to thrive in a variety of habitats. It may intermingle with mangroves on low silty land or grow vigorously on beach berms of coral and sand.

Uses

Ornamental It has been widely planted as a low-maintenance ornamental on both seaside and upland locations.

Medicinal The leaf tea has been used to quell colds, fever and edema, to aid early lactation problems, to lower blood pressure and as a lotion for skin problems. The fruit has been employed for the treatment of skin eruptions and the juice is applied to warts. Oil from the seeds is used in India to treat cutaneous infections. Latex from immature fruits has been used to treat ringworm and other itchy skin conditions. The bark tea has been used to treat skin diseases, dysentery and hemorrhoids. The juice from stems has been used to treat herpes. The yellow juice from the fruit stalk is applied externally to sprains, burns, insect bites, inflamed joints and skin diseases. Modern medicine has shown the leaves to have an antibiotic activity.

Edible The flowers have been eaten in salads, boiled and fried in a light batter. The young leaves may also be eaten raw or cooked.

Physical The attractive, strong, red and brown wood is durable and takes a fine polish. It is resistant to termites and has been used in cabinetmaking and boatbuilding. The fibrous bark has been used to make rope, for weaving, and as a source of red dye. The sap of immature fruits and buds also yields a dye. The seed oil has potential industrial use.

Ecological This is a good choice as a border tree to plant on a beach site too salty for most plants.

Notes The genus is from the Greek word meaning "marvelous." The species name refers to the plant's resembling species of the genus *Populus*. The common names haiti-haiti and portiatree are also widely used for this tree. The seeds contain 25% protein, 20% oil, beta-sitosterol, ceryl alcohol, the yellow flavone pigment thespesin, gossypol and a significant amount of phosphoric acid. The dark-red semi-drying oil with a pleasant odor is composed primarily of oleic and linoleic acids with lesser amounts of myristic, palmitic and stearic acids along with various volatile compounds. The flower petals contain the monoglucoside populin, the flavone populnetin, the phenolic compound populneol, herbacetin, kaempferol, nonacosane, lupenone, myricyl alcohol, lupeol and beta-sitosterol. The wood and flowers contain gossypol. There has been considerable work on the chemistry of this plant with less than unanimous conclusions by the various investigators.

SIDA

Sida rhombifolia   *Malvaceae*

Geographic Distribution The genus of about 200 species is found throughout the warm parts of the world but mostly in Latin America. This weedy species is now found from the southeast United States and Bermuda south throughout the Caribbean, Central America and most of South America. It has also been naturalized in the Old World tropics.

Form A small shrub to 1 m (3 ft) tall with a woody stem, deep taproot and toothed alternate rhomboid leaves. The very similar *S. acuta* has smaller and narrower lanceolate leaves and two awns on its carpels.

Flower & Fruit Date The yellow 12-mm (1/2-in) wide flowers are born at the base of leaves in all months. The flowers open by mid-morning but wilt by late afternoon.

Reproduction The pyramid-shaped black seeds develop in a capsule which splits open at maturity.

Propagation *Sida* grows from seed and is often considered to be a weed. Seed germination is inhibited by bright light and enhanced by scarification or leaching in water. When control is desired, 2,4-D strongly inhibits seedling emergence.

Habitat & Ecological Distribution This pioneer species likes full sun and is often one of the early colonizers of disturbed dry areas. Once it is established it is very competitive. On overgrazed pastures it may become the dominant weed because of its unpalatability to livestock. It is resistant to mowing and grows back promptly from its woody stumps and persistent taproot.

Uses

Ornamental This is a tough and sturdy little plant which can withstand considerable abuse. It is particularly suited to areas of periodic high foot traffic such as parking lot dividers in fairgrounds.

Medicinal In folk medicine a tea prepared from the leaves has been used to treat indigestion, gonorrhea, fevers, hemorrhoids and diarrhea. The leaves are used to promote abortion. The root tea has been used to treat

fevers, rheumatism, nervous and urinary diseases, constipation, and abdominal problems. The pulped roots are applied to sore breasts or mixed with sparrow dung to mature boils. The flowers are applied to wasp stings. Prior to WW I, a German, Dr. Dieseldorf, established a clinic to treat tuberculosis of the skin, bones and intestines  using extracts of this plant. He also claimed that regular consumption of the extract would eliminate all craving for alcohol. More recent research has shown the plant contains several pharmacologically active alkaloids. Cryptolepine and vasicine are hypotensive and antimicrobial. The combination of the stimulant ephedrine with the bronchodilator vasicinone could account for the major therapeutic use of this plant for the treatment of asthma and other chest ailments by the Indian Ayurvedic system of medicine.

Edible The leaf tea is sometimes drunk as a refreshing beverage.

Toxic The young leaves may be toxic to grazing animals and the ripe seed capsule is reported to kill poultry.

Physical The bark yields a white lustrous fiber used in cordage and textiles.

Notes The genus is an ancient Greek plant name, and the species refers to the rhomboid shape of the leaves. This plant contains three groups of alkaloids: the sympathomimetic amines beta-phenethylamine, ephedrine and psi ephedrine; the quinazolines vascine, vascicinol and vasicinone; and carboxylated tryptamines. The ratios of the alkaloids and the plant parts in which they are found vary significantly with the age of the plant. Hydrocyanic acid has been found in the bark, roots and leaves. The seeds contain about 14% oil which is composed of sterculic and malvalic acids in addition to the expected palmitic, palmitoleic, stearic, oleic and linoleic fatty acids. The seed oil has been found to contain cyclopropenoid fatty acids which have a profound effect on animals and have co-carcinogenic properties.

SEA HIBISCUS
Hibiscus tiliaceus   *Malvaceae*

Geographic Distribution The genus of about 200 herbs and shrubs is distributed worldwide in tropical and temperate zones. This species originated in the Old World and is now naturalized on tropical shores throughout the world.

Form A large shrub or spreading tree to 12 m (40 ft) tall with a dense hemispherical crown and a trunk to 25 cm (10 in) in diameter. An inland form may reach 25 m (80 ft) tall with a trunk 45 cm (18 in) in diameter. The large nearly round heart-shaped leaves, dark green above light green and downy below, may reach 15 cm (6 in) in diameter.

Flower & Fruit Date The 8-cm (3-in) funnel-shaped yellow flowers with 5 petals have a deep maroon center. They turn orange-red and usually drop by the end of the day but may persist a subsequent day. Most trees flower continuously.

Reproduction The 1-inch-long elliptic five parted seed capsules split open at maturity releasing the large black seeds. They can withstand long immersion while being dispersed by seawater.

Propagation Large branch cuttings about 25 cm (10 in) in length from the previous year's wood should be stripped of all leaves and

placed in moist soil. Treatment with rooting hormones considerably increases the rate of success. It can also be grown from seeds. It is a bit slow to start, but once it is established it grows rapidly. If left unpruned, the long lower branches are inclined to take root when they touch the soil. This tree is very frost sensitive.

Habitat & Ecological Distribution The sea hibiscus is able to grow in wet salty soils as well as dry ground. The drooping rooted branches may allow a single tree to cover a large area as a dense thicket.

Uses

Ornamental It is grown as an ornamental for the showy flowers and lush foliage. It should be pruned at the base when young to produce a clear trunk if it is destined to be a shade tree. The long flexible branches allow it to be trained in an arch.

Medicinal A tea prepared from the bark has been used to treat fevers, bronchitis, asthma and to encourage the growth of hair when applied to the head. The bark macerated in water makes a mucilaginous solution used to treat dysentery. A tea made from the flowers and root bark is used to relieve colitis, indigestion and excess stomach acidity. The tea prepared from the seeds is emetic. The leaves are used as a laxative, emollient and cure for ulcers. An acetone extract of the leaves has been shown to have antibacterial effects against *Staphlococcus aureus.*

Edible The tender inner bark and young leaves may be eaten. The flowers may be eaten raw in a salad, boiled or fried in a batter.   The roots are valued as food by aborigines.

Physical The wood is durable in salt water and has been used as floats for fishing nets and outrigger canoes, light boat planking, and pilings for houses in swampy spots. It is a choice wood for primitive fire starting by friction. The tree is inclined to send up many long vigorous sprouts from cut stumps. The fibrous inner bark from young sprouts has been used for heavy cables, ropes, harpoon lines, fish nets, fish traps, mats, coarse cloth and is the traditional material for hula skirts. The bark fiber is stronger wet than dry.

Ecological The leaning trunks and rooted branches aid in stabilizing dunes and muddy shorelines. This tree is a useful component of a mixed species coastal windbreak planting.

Notes The genus is from the Greek name for the related marsh mallow. The species is from the Latin name of the linden tree whose foliage it resembles. The name *majo*, derived from an American Indian name applied to several trees with a fibrous bark, is also commonly used. This tree is the source of nectar for honey. The vigorous growth habit has caused it to be considered a pest in some areas. The flowers contain flavanoids and glycosides of gossypetin along with quercetin and kaempferol. The fruits contain beta-sitosterol, quercetin, kaempferol, fumaric acid and p-coumaric acid.

PITCH APPLE
Clusia rosea   *Guttiferae*

Geographic Distribution The genus of 145 species is found primarily in the
   tropical Americas. The pitch apple occurs throughout the Bahamas and
   West Indies and has been extensively introduced in Florida.

Form A tree with a dense broad crown, it grows to a maximum of 18 m (60
   ft) in height with a 60-cm (2-ft) trunk diameter. The pitch apple usually
   begins life on the branch of another tree, sending aerial roots to the soil
   below. The rapidly growing roots form a trunk and often strangle the
   supporting tree. When directly planted in the soil on a sunny site, this tree
   grows about as broad as tall. The large thick opposite leaves are smooth
   and glossy with a sturdy petiole.

Flower & Fruit Date The showy white 8-cm (3-in) flowers, highlighted with
   pink and yellow, may occur in any
   month of the year.

Reproduction The attractive light-green
   globular segmented fruits, 5 cm (2 in)
   in diameter, turn brown at maturity and
   split open to release the many seeds.

Propagation The tree is easily propagated
   by cuttings and air layers. Natural reproduction is almost always as an epiphyte from seeds dropped and lodged
   in branch crevices by birds. Fresh seeds
   germinate well after about 30 days and
   grow at a modest rate when planted in
   a standard potting mix and maintained
   at a daytime temperature between 85°
   and 95°. This temperature requirement
   explains why so many people in Flor-

   ida comment that if you plant the seeds in fall, winter or spring, they will
   not germinate until June or July. Growth rate of the established tree is
   moderate to rapid depending on the soil quality. It should only be planted
   in frost-free areas.

Habitat & Ecological Distribution The tree is tolerant of high salt content
   in the soil and can withstand constant wind but needs full sun. It may be
   found on rocky sea cliffs just above the reach of waves with no apparent
   soil, and inland to higher elevation forests.

Uses

   Ornamental The large attractive downward-looking flowers and attractive leathery leaves combined with wind and salt tolerance make the pitch
   apple a desirable ornamental for exposed seaside planting. It may be
   pruned to shape as a hedge, espalier or topiary or left to develop as a

symmetrical shade tree. It is commonly planted around shopping plazas and similar stressful public environments. The aggressive roots may be destructive. A commercially propagated clone "*Variegata*" with marbled leaves is increasingly popular.

Medicinal The yellow latex has been used as a balm on the skin to heal various ailments and has been applied to cavities to relieve toothache. A tea made from the leaves or flowers has been used to aid chest problems,

and a tea from the fruit rind or bark has been used to treat rheumatism. The dry powdered latex is sold commercially in South America to be used as a plaster to promote healing of fractures, dislocations and burns. The resin contains xanthochymol. *Toxic* The latex is dangerously purgative if taken internally.

Physical The strong reddish-brown wood is used as fuel and for fenceposts and crossties. The viscous yellow resinous latex hardens and turns black on exposure to air and has been used as boat caulking and plaster.

Ecological Many species of birds enjoy the seeds after the fruit splits open.

Notes The genus is named for the 16th-century French botanist Lecluse. The species name probably refers to the rosy tints of the flower. Fat pork, cupey, wild mammee and autograph tree are other common names. The name autograph tree derives from the fact that a mark scratched on the surface of the leaf produces a scar which is evident on the leaf even after it has fallen from the tree. The early conquistadors used the leaves as writing paper and playing cards by marking the surface with a pin. The resin contains aplotaxene, friedelin, A and B friedenols, oleanic acid and sitosterol.

WILD CINNAMON
Canella winterana   *Canellaceae*

Geographic Distribution The genus of 2 species is found throughout tropical America. This species is found from South Florida through the Greater and Lesser Antilles south to Barbados and has been introduced into northern South America.

Form This small tree with an open canopy reaches a height of 9 m (30 ft) with a diameter of 20 cm (8 in). The spatulate, glossy dark-green leaves grow in clusters at the ends of twigs.

Flower & Fruit Date The 6-mm (1/4-in) dark purplish-red flowers in clusters at branch tips may be found in any month of the year but are most abundant in the winter.

Reproduction The spherical red or purplish-black fleshy berries usually found in the spring have several black seeds.

Propagation Seeds fresh from the tree and cleaned of fruit pulp before planting germinate best. The root system needs room for an initial burst of growth so the young plants should be set out in their permanent location when only a few inches tall.

Habitat & Ecological Distribution This tree may be found growing on dry sandy or rocky shorelines from the high-tide line inland in dry forests at low elevations. It is tolerant of salt and drought. It does best in full sun but tolerates open shade easily.

Uses

Ornamental This small tree is planted as a hardy ornamental and for its numerous colorful red berries.

Medicinal The bark was formerly harvested and sold for pharmaceutical use in the U.S. and Europe. As a home remedy a tea prepared from the inner bark has been used to treat fever and indigestion, as a gargle for sore tonsils, as a tonic and to treat menstrual disorders. The leaves are also used to relieve aches and pains. Canellal (also called muzigadial), a sesquiterpene dialdehyde isolated from the bark, is antimicrobial, antifungal, cytotoxic and inhibits insect feeding.

Edible The aromatic inner bark is not the cinnamon of commerce although it has a similar scent due in part to the volatile oils eugeneol and cineole. It is sometimes called pepper cinnamon, white cinnamon or wild cinnamon. Within the native range of the tree the bark has been used directly as a spice and is exported for use as a component in spice blends. It is used as a flavoring agent in several proprietary products and in some aromatic smoking tobaccos. When gathered green and

dried, the berries are also used as a spice.

Toxic The bark is insecticidal and has been used to poison fish. The stems and leaves are toxic to chickens. The stems are reported to be repellent to cockroaches.

Physical The heartwood is almost black and the sapwood is greenish brown. The wood is very strong, hard and sometimes heavy enough to sink in fresh water. It is used as posts, poles, beams and plow frames. It takes a smooth finish and a high polish. The volatile oils have been used in blending perfumes.

Ecological The flowers produce abundant nectar and are heavily utilized by bees.

Notes This tree has also been called *Canella alba* and *Winterana canella.* The genus *Canella* is Latin for "cinnamon" and the species *winterana* honors captain Winter who intoduced the plant to Europe. Other common names used are barbasco, cinnamon bark, caneel, and canella. The bark contains 1% volatile oil containing pinene, eugenol, cineole, caryophyllene resin, mannitol, helicid, clovanidiol, myristicin, canellal, warburganal and several drimane sesquiterpenes. The leaves contain epoxymuzigadial, a derivative of the muzagadial found in the bark.

YELLOW ALDER
Turnera ulmifolia   *Turneraceae*

Geographic Distribution The genus of about 60 species is found in the American tropics and subtropics. This species is found in Bermuda, Florida, the Bahamas and tropical America. It has been introduced in tropical Africa and Asia as well as islands in the Indian Ocean.

Form A sprawling or erect herbaceous shrub up to 3 m (10 ft) but usually less than 1 m tall. The alternate lanceolate leaves are coarsely toothed and up to 14 cm (5 in) long by 5 cm (2 in) wide.

Flower & Fruit Date The bright yellow flowers with 5 petals to about 5 cm across open in the morning and secrete a nectar containing sucrose in twice the concentration of glucose and fructose. The flowers are incompatible to their own pollen and are visited by numbers of bees of the genus *Apis* and *Melipona.* The flowers last only one day and begin to wilt in the early afternoon. Flowers are borne throughout the year.

Reproduction The three-parted pod contains oblong 2.5-mm (0.1-in) curved seeds.

Propagation This plant grows well from either seeds or cuttings.

Habitat & Ecological Distribution May be found on sandy dunes, rocky shorelines or inland in patches of sunshine in coastal forests.

Uses

Ornamental This small shrub with abundant flowers makes a good addition to a varied border planting or a flower bed.

Medicinal A tea prepared from the foliage has been employed to treat menstrual disorders, colds, diarrhea, hemorrhoids, bronchitis, constipation, eyestrain and toothache. The pounded leaves have been applied externally to treat sores. Another member of this genus is reputed to have potent aphrodisiac properties. The seeds contain up to 1.2% caffeine and are a potential source of this drug.

Toxic Extracts of the foliage contain cyanogenic compounds and are toxic to mosquito larvae. They also show synergistic toxic effects when combined with various synthetic larvicides.

Notes The genus is named for William Turner, an herbalist who is considered to be the father of English botany. The species *ulmifolia* refers to the leaves resembling elm leaves. In addition to the nectar and pollen produced by flowers, this plant has a nectar-secreting structure in the foliage at the base of its leaves. The glands produce a nectar with a balanced combination of sucrose, glucose and fructose which increases in concentration through the day due to evaporation of the water content. The foliar nectaries have probably evolved to attract ants to guard against herbivores which might damage the plant. Members of this family produce cyclopentanoid cyanohydrine glucosides in their foliage. In this species the amino acid cyclopentenylglycine has been found to be a precursor of the cyanogenic glycoside deidaclin.

WOOLLY NIPPLE CACTUS
Mammillaria nivosa   *Cactaceae*

Geographic Distribution While the genus has several hundred species in the American tropics, this species is the only native one in our area. It is found in the Bahamas, the islands around Puerto Rico and in the Lesser Antilles.

Form A slightly flattened spherical cactus to 10 cm (4 in) in diameter with no ribs but numerous tubercles each bearing up to 10 spines and producing white wool. Often growing in clumps.

Flower & Fruit Date The 9-mm (3/8-in) cream flowers are borne on the tubercles.

Reproduction The bright red oval berries up to 12 mm (1/2 in) long contain small brown seeds.

Propagation Only the smallest seedlings should be transplanted from the wild as the larger specimens seldom survive the experience. The seeds can be planted on a light potting soil and covered with a sprinkling of sand. With restrained watering and considerable patience they will produce many young plants. With all cacti, the pot with seeds may be covered with glass or plastic to maintain a constant modest humidity. Take care not to keep the seeds or seedlings too moist. Under conditions of excessive moisture this cactus is susceptible to Fusarium rot. If kept in direct sun, covers of germination pots should be  somewhat elevated so that air circulation can prevent overheating. Suckers produced at the base of adult plants may be carefully transplanted.

Habitat & Ecological Distribution This cactus is most frequently found on open sites on dry rocky shorelines.

Uses

Ornamental The fuzzy little balls formed by this cactus make it an attractive addition to a xeriscape or rock garden. The only care they require is periodic weeding to keep other plants from shading them.

Edible The berries are edible.

Ecological On small rocky islands or in very parched terrain the juicy berries from this cactus provide an important source of fresh water for many birds and lizards.

Notes The genus is from the Latin word referring to the nipplelike tubercles and the breastlike growth form. The species is from the Latin meaning "snow," referring to the frosted appearance produced by the white wool. This has also been called snow cactus.

T URK'S CAP CACTUS

Melocactus intortus   *Cactaceae*

Geographic Distribution The genus of about 36 species is found in the tropical Americas. This species is found on dry islands in the Bahamas, Puerto Rico and the Lesser Antilles.

Form An upright oval barrel cactus rarely to 1 m (3 ft) tall with 15 to 20 ribs.

Flower & Fruit Date Mature individuals develop a woolly vertical terminal structure called a cephalium. This species has a red bristly cephalium which takes several years to develop to its full size of 8 cm (3 in) in diameter by up to 30 cm (12 in) tall. The bright pink flowers occur between the bristles at the upper end of the cephalium.

Reproduction The hot pink, glossy, elongate, cone-shaped fruits are borne among the wool and bristles of the cephalium. Each fruit containing many small shiny black seeds is ejected when ripe.

Propagation The extensive shallow root system makes transplant from the wild difficult. Very small individuals may be transplanted successfully if gently cared for in their new setting until established. New plants may be started from seed planted shallowly in light soil. Once the cephalium begins to grow, the body of the cactus ceases all growth, becomes more woody and develops a corky cambium. Damage to the upper body of the cactus usually results in the production of additional small heads, each of which produces a cephalium. If grown as a houseplant in more northern regions it should be given at least a half day of direct sunlight daily and should not be allowed to chill below 60° F. When preparing soil for potting or a planting site, the inclusion of some carbonate sand or ground limestone will maintain the pH at a suitable level. Water can be applied once a month when there is no rain, and fertilizer can be lightly applied once a year.

Habitat & Ecological Distribution This cactus is found on open rocky dry shorelines. It is most abundant in areas so dry that its competitors cannot grow over it to shade it.

Uses

Ornamental Widely planted and preserved as an ornamental on dry rocky sites.
Edible The sweet juicy fruit is edible and generally appreciated in the hot dry cactus habitat. A handful of the bright fruit adds interest to fruit or vegetable salads.
Toxic Skin punctured by the spines seems to be prone to inflammation and festering.
Physical The body of the cactus has been sliced and used as bait in fish traps.
Ecological The plant stabilizes the soil with an extensive root system and the fruits are eagerly consumed by many species of birds, anoles and iguanas.

Notes The genus name refers to the irregular spherical shape and translates as "apple shaped melon." The species name means "twisted upon itself." The common name refers to the resemblance of the cephalium to a fez, the red felt hat, shaped like a frustum of a cone, worn by Turks and other eastern Mediterranean men.

PRICKLY PEAR
Opuntia dillenii   *Cactaceae*

Geographic Distribution There are many species of *Opuntia* in the Western Hemisphere. This is one of the most common and is found from Bermuda south to Florida, the Bahamas, and throughout the West Indies. It has been introduced and naturalized in Africa, Asia and Australia.

Form A spreading low-growing cactus to 2 m (6 ft) in height. The fleshy pads may be up to 30 cm (12 in) long.

Flower & Fruit Date The flowers vary from yellow to orange, sometimes with a tinge of red at the base. They are up to 5 cm (2 in) across, and may be found in any month.

Reproduction The pear-shaped stemless fruit contains many seeds, takes about 4 months to mature and is deep red to purple when ripe.

Propagation Vegetative propagation is most successful. Some fruits develop flowers at their ends. When these are detached and placed on soil, they send down roots and develop vegetatively into new plants. A joint or pad dried for a few days, then placed on the soil, will also take root and produce a new plant vegetatively. Germination from seed is possible but more difficult and time consuming. When grown as a houseplant it needs at least a half day of direct sunlight. As a yard plant, water and fertilizer are not needed.

Habitat & Ecological Distribution Found on beaches, dunes and dry rocky coastlines. They do best in full sun on a well drained soil but can tolerate an open shaded location.

Uses

Ornamental This cactus makes a hardy addition to a rock garden or xeriscape. The many related species and cultivars with red, purple, yellow and white flowers are being used more frequently as integral landscape components or as a distinctive focal point.

Medicinal A tea prepared from the pads has been used to treat inflammation and ulcers. The flesh of the heated or cooked pad has been used to treat boils, splinters, tumors, pleurisy, arthritis and rheumatism and for extracting guinea worms. The joint or root has been crushed in water and the resultant beverage consumed to relieve urinary burning. The flowers have been used to treat diarrhea (the effectiveness of the treatment is probably due to the presence of a flavonoside). The ripe fruit is eaten to treat gonorrhea,

and baked it is used to treat asthma, liver congestion and whooping cough.
Edible The ripe fruit, often called tuna, should be picked with thick gloves as protection from the bunches of small spines on the surface. Ripe fruits are edible raw, sliced into a fruit salad or baked with cinnamon and butter. They can also be cooked into jams and jellies, or fermented into wine. Consumption of large amounts of the fruit is said to produce constipation. The young pads, called nopalitos, with soft immature spines, can be used as a component of a dip, diced and included in 3-bean salad, peeled and marinated then added to a green salad, battered and fried, and used in stir-fry after being diced or shredded. The older pads can be despined and roasted or boiled in a soup. In dry times cattle and burros are able to eat the pads with no visible adverse effects.
Physical The juice extracted from the pads has been used as a shampoo. Both mucilages and pectins have been found in the pads. The fruit juice has been used as a red ink.
Ecological The seeds and pulp of the fruit are consumed by many birds and mammals.

Notes The genus is derived from the name of a plant which grew near the ancient Greek town Opuntis. The species is named to honor J. J. Dillenius, an early 18th-century British botanist. This cactus is also known as *O. stricta*, and "tuna" is also used as the common name. The larvae of a moth (*Cactoblastis cactorum*) may devastate colonies of this cactus. This moth has been used in Australia to help control pestiferous populations of this cactus. A red, purple or orange dye called cochineal is derived from another insect which lives on this cactus. A 100-g (4-oz) sample of the fruit contains: 42 calories, 88 g water, 0.5 g protein, 0.1 g fat, 10.9 g carbohydrate, 20 mg calcium, 28 mg phosphorus, 0.3 mg iron, 2 mg sodium, 166 mg potassium, 60 I.U. vitamin A, 0.003 mg thiamine, 0.014 mg riboflavin, 0.32 mg niacin and 22 mg vitamin C. The 15% sugar content of the fruit is about equally divided between glucose and fructose. The pigment betanin provides the color. An arabogalactan has been isolated from the pads. The mucilages and pectins from the stems are composed of glucouronic and galactouronic acids, glucose, galactose, arabinose and xylose polysacharides. The fruit contains glucose and fructose but almost none of the more common plant sugar sucrose. A yeast *Saccharomyces opuntiae* found on the ripe fruits ferments glucose and fructose but not sucrose.

Dildo cactus
Pilosocereus royenii　　　　*Cactaceae*

Geographic Distribution The genus of about 40 species is found from Mexico through the West Indies to Brazil. This species is found from Puerto Rico to Antigua.

Form A tree cactus to 6 m (20 ft) tall with a trunk of 30 cm (12 in) in diameter, branching a few feet off the ground to form columnar branches with 7 to 11 ridges. The thick green skin on the branches carries out photosynthesis. The root system is broad and near the surface to rapidly absorb water from even the lightest rains.

Flower & Fruit Date The white to purple flowers are borne directly on the branches when rainfall is adequate in any month.

Reproduction The 2- by 5-cm (1- by 2-in) red berries with many small black seeds grow directly on the branches.

Propagation The seeds may be germinated and grown in containers, or young plants may be transplanted from the wild. Young specimens do well as pot plants if given at least 5 hours of direct sunlight each day.

Habitat & Ecological Distribution This cactus may be found in almost any dry site from beach berms to rocky cliffs. This cactus grows slowly on harsh sites. A 5-m (15-ft) tall individual may be over 100 years old.

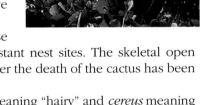

Uses

Ornamental The striking candelabra growth form and minimal care required make this an excellent landscape accent plant on dry sites.

Edible The sweet fruits may be eaten fresh, made into a beverage or preserved as a condiment.

Physical Chunks of the branches have been used as bait in fish traps.

Ecological Doves and other birds use the branch crotches as predator-resistant nest sites. The skeletal open network of wood fibers remaining after the death of the cactus has been used in various craft work.

Notes The genus is from the Latin *pilo* meaning "hairy" and *cereus* meaning "candle shaped." This cactus is also known as *Cephalocereus royenii*.

RED MANGROVE

Rhizophora mangle   *Rhizophoraceae*

Geographic Distribution The genus of 7 species is found throughout the tropics. This species is found in West Africa, on both coasts of the tropical Americas and westward in the Pacific to Polynesia and Melanesia.

Form From a small shrub to a large tree, the red mangrove usually grows in thickets rendered impenetrable by the many interdigitating arched and branched prop roots descending from the trunk, branches and other anchored aerial roots. The prop roots contain numerous above ground lenticels which provide oxygen for the roots immersed in anoxic mud. The roots are able to reject the excess salt from the environment, leaving the sap with a salinity typical of other plants. The tall straight trunks may extend to 20 m (70 ft) or more and reach a diameter of 1 m (3 ft).

Flower & Fruit Date The four-petaled yellow flowers are pollinated by the wind and occur throughout the year.

Reproduction The single seed germinates on the tree and produces a thick pendulous root resembling a giant bean up to a foot in length before dropping into the mud to produce another tree. These propagules float and survive suspension in seawater over a year or stranding on a dry beach for two months; thus, winds and currents disperse them to colonize all suitable habitat. The propagules can develop while lying on their side or when vertically embedded in the mud. Rooting is equally rapid in fresh or salt water. This is the only mangrove which can colonize new habitats below low tide.

Propagation The rooted seeds (hypocotyls) sprout and grow vigorously in moist mud. Larger trees may be transplanted successfully if a shallow root ball at least the diameter of 1/2 of the tree's height is taken up with the tree. A vigorous pruning restricted to branches of less than 12 mm (1/2 in) in diameter aids the establishment of transplants. Long-term clipping and pruning of plants may result in their death.

Habitat & Ecological Distribution This tree is characteristic of shallow protected muddy coastlines where the broadly spreading stilt roots allow the trees to advance seaward in water as much as 1 m (3 ft) deep. Red mangroves can tolerate salinities from fresh water to 44 parts per thousand. The maximum height of the trees in a mature stand is inversely proportional to the soil salinity in the range of 17 to 72 ppt. Trees begin to die when soil salinities exceed 65 ppt. and all are dead if the salinity reaches 90 ppt. Thus rainy periods allow red mangroves to invade new habitat, while droughts increase salinity and kill red mangroves which have colonized salt flats and similar areas. Soils formed under mangroves are typically anaerobic with a high pH. Upon drying, oxidization results in intense acidity rendering the soils unsuitable for agriculture. Freezing temperatures are lethal, but water temperatures of 104° F are tolerated. They are consistently able to withstand hurricane

winds with little obvious structural damage. Dormant buds live only 3 years; thus a tree stripped of its young branches by a storm will not be able to send up new epicormic shoots and will die after it has served as a buffer to storm winds and waves. Young trees require full sun and will not grow under a closed canopy stand.

Uses

Ornamental The red mangrove forms a rich dark-green border or privacy screen for property fronting on shallow tidal waters.

Medicinal Powdered leaves have been used to relieve constipation, and a tea from the leaves has been used to treat fish poisoning. The astringent nature of the bark (it contains 15% to 40% tannin) has led to its use in treating hemorrhage, inflammation, diarrhea, dysentery, leprosy, elephantiasis, syphilis and gonorrhea.

Edible A tea has been made of the dried leaves, and they have been used as a tobacco substitute. The bitter astringent fruit is suitable only as emergency fare. Fresh mangrove leaves have been suggested as a supplemental poultry and cattle feed due to their high nutritional value.

Physical The reddish to purplish-brown strong, elastic, hard, heavy wood has been used in boatbuilding as ribs, decks and knees, also for oars, wharves and pilings. On land it is used for rafters, joists, window frames and furniture. The wood finishes and polishes well, developing its dark color over time as does mahogany. It is suitable as a raw material for particle board. It is resistant to termites and decay but not marine borers. The bark yields 18% by volume of a pyrocatechol-type tannin. It is commercially important in tanning where it imparts a dark red color to the leather. A bark extract is used for the preparation of phenolic adhesives used in plywood manufacture. A red dye from the shoots is used on leather. Other dyes for staining wood and textiles may be made by treating extracts of the bark and roots with iron or copper compounds to produce brown, slate, rust and olive colors. Paper produced from the wood is of a low quality unless longer fiber from other species is added to the pulp. The wood makes an excellent charcoal and burns well even when freshly cut.

Ecological The dense coppice of prop roots quells water motion and serves to prevent wave-induced erosion. The maze of small convoluted channels produced by the mature mangrove stand slows terrestrial runoff and promotes sediment retention. The abundant substrate and rich supply of organic detritus derived from the shed leaves provide an essential nursery for the immature forms of many marine organisms including commercially valuable species such as crabs, spiny lobsters and snappers.

Notes The genus is from the Latin meaning "to bear roots," referring to the conspicuous prop roots. The species *mangle* is derived from the Arawak Indian word for mangrove. The roots serve as a substrate for the delicious and much-sought-after mangrove oyster (*Crassostrea rhizophorae*), known as the oyster which grows on trees. The buffeting of hurricane winds seems to cause internal damage leading to the death of entire forests of mature intact trees. Oil spills also threaten the trees by coating the roots and clogging the lenticels which provide essential oxygen for root metabolism. The leaves have a potential as a component of livestock feed. The nutritional content of 100 g (4 oz) of mangrove leaves is: 7.5 g protein, 3.6 g fat, 1.3 g calcium, 0.14 g phosphorus, 0.65 g potassium, 30 mg magnesium, 15.2 mg iron, 3.5 mg copper, 0.52 mg cobalt, 4.3 mg zinc, 54 mg iodine, 0.6 mg vitamin A, 13 mg vitamin B1, 19 mg vitamin B2 and 240 mg niacin.

BUTTONWOOD

Conocarpus erectus　　　*Combretaceae*

Geographic Distribution The two species in the genus are found throughout the warm parts of the world. The buttonwood is found on both coasts of tropical America and in western tropical Africa. It has been introduced and naturalized in Hawaii.

Form A low shrubby tree with a spreading crown sometimes growing to 20 m (60 ft) feet in height with a trunk of 75 cm (30 in). The alternate leathery leaves are lance shaped with 2 salt glands at the base.

Flower & Fruit Date The minute greenish flowers occur as fuzzy balls intermittently throughout the year.

Reproduction The brown spherical conelike fruits remain on the tree for long periods of time before the scalelike seeds separate and fall to the ground. Unlike the true mangroves, the seeds never germinate and develop on the tree.

Propagation The germination rate of seeds is low. Vegetative propagation from rooted cuttings is more reliable. Growth rate is generally slow but the green form grows faster than the silver-haired variety.

Habitat & Ecological Distribution Generally found near salt water from muddy swamps to dry rocky shorelines. It cannot tolerate sites as wet as those occupied by white and black mangroves, but it can also survive harsh periods of drought. Thus it is usually the most shoreward of the trees in mangrove swamps and may occur at some distance from the water.

Uses

Ornamental The buttonwood is becoming widely used for landscaping as a bordering tree, screen and as a clipped hedge. Silky silvery hairs on the foliage of one variety produce an attractive silvery-appearing plant. The more abundant dark-green form is more cold resistant.

Medicinal A tea prepared from the bark is used externally to treat prickly heat and inflamed eyes. Internally the tea is drunk to treat syphilis and diabetes.

Physical The leaves and bark have been used in tanning. The light-brown to gray wood is hard, heavy and takes a fine polish. It has been used in marine construction and is often used as ribs in native boats. It is a favored wood for smoking fish and makes a high-quality charcoal.

Ecological The tree's tolerance of drought, salt and strong wind make it an ideal windbreak for windward shores. It provides habitat for *Anolis* lizards, the presence of which reduces plant damage by insects.

Notes The generic name *Conocarpus* refers to the conelike fruits, and the species name *erectus* refers to the upright growth of the seedlings. The name button mangrove is also used for this tree although it lacks most of the characteristics of a true mangrove. The repeated use of button in the common names refers to the resemblance of the fruit to old-fashioned shoe buttons. The silvery-haired form (*sericea*) is a reoccurring mutation or genetically recessive and is not reliably transmitted by seed so it must be propagated by vegetative means. The bark of both forms has 16% to 18% tannin.

WHITE MANGROVE
Laguncularia racemosa   *Combretaceae*

Geographic Distribution This single species genus is found throughout the Caribbean, both coasts of tropical America, and tropical West Africa from Senegal to Angola.

Form An evergreen tree to 25 m (80 ft) tall and a trunk to 70 cm (30 in) in diameter, frequently with multiple trunks. In soft muddy soils the trees often send out an above-ground cone of prop roots. They may also send

up vertical pneumato-phores to allow the horizontal root system to breathe in saturated soils. The leathery elliptic leaves are opposite and have red petioles with two salt-ex-creting glands near the leaf blade.

Flower & Fruit Date The small, white, fragrant, bell-shaped flowers are borne in clusters at the ends of branches throughout the year. They are pollinated by insects.

Reproduction The pear-shaped, slightly flattened, ribbed, green-gray fruits with a single seed germinate and begin seedling development while still on the tree. The young plants (seeds) begin to lose vigor after 8 days but can survive floating in the sea for over 30 days and are broadly dispersed by winds and currents. When the seed sprouts, it first extends a root. It must then be stranded above the influence of tidal disturbance until the root penetrates the substrate. The seed is then pulled into a vertical position before the stem elongates and the leaves emerge. Development to the point of resisting tidal dislodgement is rapid.

Propagation Fresh seeds germinate and grow readily in moist soil. Air layers produce adequate roots after 6 months. Naturally growing seedlings may be transplanted bare root, and larger trees may be transplanted with a root ball at least 1/2 the diameter of the tree's height. Vigorous pruning of

branches to 5 cm (2 in) in diameter aids survival and recovery. A balanced fertilizer accelerates growth rate. A mulch of seagrass gathered from the nearby beach aids the growth of young plants.

Habitat & Ecological Distribution The white mangrove prefers moist silty soil. It can grow in seawater-saturated soil if the site is shallow andprotected from strong wave action. As soil salinity begins to exceed that of seawater the growth rate is reduced and it is often replaced by black mangrove. This is the most cold-sensitive of the mangroves. The shallow root system makes this tree susceptible to being tipped over by strong winds.

Uses

Ornamental The white mangrove can be used as a trimmed or natural hedge and border plant on waterfront sites.

Medicinal The bark tea has been used as a tonic and to treat dysentery. A bark extract has been found to have antitumor activity.

Physical The hard, heavy, strong wood has been used for tool handles, fuel, house framing and posts. It is resistant to termites but heavily damaged by marine boring organisms. The bark contains both pyrocatechol and pyrogallol tannins. The bark and leaves have been used in tanning and dyeing.

Ecological Growing on the interface between land and sea, a stand of these trees with their dense network of roots helps reduce sediment in water runoff from land and helps reduce coastal erosion from storms. The root structure and growth habit make it especially suitable for controlling erosion on canal banks. The shed leaves contribute to the fertility and function of the adjacent marine ecosystem.

Notes The genus *Laguncularia* is from the Latin *laguncula* and refers to the resemblance of the fruit to the small hand-blown glass bottles of the day. The species *racemosa* is from the Latin describing the racemes or clusters of flowers. The bark contains 10% tannin and the leaves 17%. Over a million kilograms of white mangrove leaves per year are used in Brazilian tanneries. It is a honey-producing plant.

INDIAN ALMOND

Terminalia catappa   *Combretaceae*

Geographic Distribution The genus has about 200 species spread worldwide in the tropics. A native of coastal areas from eastern India to Australia, the Indian almond has been introduced and naturalized throughout the tropics.

Form An evergreen tree to 25 m (80 ft) tall with a sometimes buttressed trunk 1 m (3 ft) in diameter. The tapered crown is characterized by horizontal tiers of branches at different levels on the trunk. The large glossy dark-green leathery leaves are alternate and crowded in rosettes at the ends of twigs, turning bright red or yellow before they drop. Deep tap roots normally develop in sand, but a more shallow lateral root system develops when the water table is near the surface.

Flower & Fruit Date The tiny greenish-white flowers are found on spikes at the terminus of leaf clusters throughout the year except for a brief period after it has shed its leaves and not regrown new ones.

Reproduction The flattened, pointed, oval fruits have a firm fleshy exterior over a fibrous husk which encloses a hard nut protecting an oily seed similar to a true almond. When mature the fruits are about 5 cm (2 in) long and turn yellow with a reddish tinge when ripe. They may fruit continuously or provide several crops per year. The fruits retain their viability for over a year. Fruit bats are major sources of fruit dispersal through much of the tree's range. The seeds are specialized for flotation with a corky rind and many tiny air cavities in the stone. They can survive floating for long periods in seawater.

Propagation The seeds germinate readily in a moist sandy soil even if the nut is not cracked. Seedlings can be grown in full sun, but do better if shaded until they are planted out. Because of the extensive root growth they should be kept in pots that are relatively large for the size of the plant and set out in their permanent location by the time they are 18 inches tall. The tree grows rapidly once it is established, reaching an average of 16 m (45 ft) in 9 years and may live for 60 years. The tree prefers full sun and a light soil but still does well in open shade and clay or rocky soils particularly if provided with some fertilizer.

Habitat & Ecological Distribution This tree is a characteristic strand plant in the Asian tropics but is now widely established on tropical American shores. It grows best in moist tropical climates but can survive with only 75 cm (2 ft) of rain and occasional light frost. While it thrives on many upland sites, its tolerance of salty, acid or alkaline soil and its wind resistance make it an ideal beachfront tree. Heavy pruning or air pollution may kill adult trees. Good drainage is required when it grows on clay soils. It is very resistant to damage from hurricane winds.

Uses

Ornamental It is widely planted as an ornamental because of its tiered pagoda shape and the brilliant color of the foliage prior to leaf drop. It is also planted as a street and shade tree.

Medicinal The leaf tea with abundant astringent tannin has been used as a tonic and to treat diarrhea, and is considered to be sudorific. The bark tea, sometimes including some fruit, has been used to treat diarrhea, dysentery, gonorrhea, asthma, bilious fever, stomach cramps and is considered to be diuretic and cardiotonic. Externally the leaf or bark teas have been used to treat skin rashes, eruptions, scabies, leprosy, and to heal the cracked nipples of nursing mothers. The young leaves have been eaten to cure headache. Traditional Ayurvedic medicine considers the fruits to be antibilious, antibronchitic, aphrodisiac and astringent. The leaves and stems

112

have been experimentally found to have antibiotic effects.

Edible The outer layer of the fruit is edible and may contain as much as 3.6% ascorbic acid. The color, shape, size and taste of the fruit vary considerably between trees. There seems to have been some informal selection for large tasty cultivars but none have been formally named or distributed. If the nut is cracked (thoroughly dried fruits are easier to crack), the seed kernel, which is similar to filberts or almonds, may be eaten raw or roasted.   The seed contains up to 52% oil known variously as talisay or Indian almond oil, which may be extracted for cooking, pharmaceutical and other uses. The remaining seed cake is a good food for pigs. The 27% protein content of the kernel also makes it nutritionally desirable, particularly in lesser developed countries where protein deficiencies are common. The kernels are commonly eaten out of hand and may be substituted for almonds in most recipes. The quality of the kernel varies considerably between trees.

Toxic Bloody diarrhea has been reported in small children as a result of eating the flesh of the fruit.

Physical The attractive reddish-brown, hard, strong, elastic wood takes a lustrous polish but is difficult to work. The wood grain is variously interlocked and sometimes curly or twisted. It has been used in boatbuilding, to carve into gongs and drums, in house construction, furniture and decorative veneer. It is susceptible to termites. The roots, bark and leaves are high in tannin, and the unripe fruit (collected, dried and marketed to tanneries as myrobalans) has up to 20% tannin. The wood, leaves, and bark have been used as a source of several dyes which can produce light brownish-yellow, golden-fawn and slate colors in silk, cotton and wool. A black dye and an ink has been prepared from the bark, fruits and foliage. The seed oil is suitable for making soap and other industrial uses but is not fully utilized due to difficulties of extraction. A gum from the tree is used as a dye source, an ink and in cosmetics.

Ecological Fruit bats are very fond of the fruit and may carry it long distances from the tree before chewing the outer flesh and dropping the intact nut. Various species of parrots feed on the ripe fruit. Fallen fruits are eagerly consumed by rodents, land crabs and various large ungulates.

Notes The genus name refers to the leaves clustered at the terminus of the twigs. The species *catappa* is a crude Latinization of the native Malayan name "ketapang." This tree is also known as *almendra* and a variety of similar names related to almond in English and Spanish. In addition to the protein and oil the seeds contain the following nutrients per 100 g; 17 g carbohydrate, 14 g fiber, 497 mg calcium, 957 mg magnesium, 9.2 mg iron, 70 mg sodium, 784 mg potassium, 0.71 mg thiamine, 0.28 mg riboflavin and 0.7 mg niacin. The amino acids present in the seed kernel include arginine, cystine, histidine, isoleucine, leucin, lysine, aspartic acid, glutamic acid, alanine, glycine, proline, methionine, threonine, phenyla-lanine, tryptophane, valine and tyrosine. The oil contains 55% palmitic, 23% oleic, 1.5% myristic, 6% stearic, and 7% linoleic fatty acids. The outer flesh contains 4.9% protein and .21% vitamin C in addition to significant amounts of other vitamins and minerals. The shell contains 25% pentosans and is a good source for making furfural. The tree yields an insoluble vegetable gum. The leaves contain steroids, diterpenes, triterpenes, flavonoides, phenolic compounds and catechic tannins. The bark and wood contain gallic acid, catechin, epicatechin, ellagic acid and leucocyanidin. The foliage of this tree is used to raise the tasar or katkura silkworm (*Antheroea paphia*). This tree has a great potential for genetic selection and development as a source of timber yielding larger fruits with a palatable flesh and large kernel. This tree has the potential of being a major resource in lesser developed countries and on sites with otherwise restrictive salinities.

TORCHWOOD
Jacquinia arborea   *Theophrastaceae*

Geographic Distribution The genus of about 35 species is found in the American tropics. This species is found from Cuba through the Greater and Lesser Antilles to Tobago. The very similar species *J. keyensis* is found in southern Florida.

Form A shrub or small tree to 5 m (15 ft) tall with a trunk to 15 cm (6 in) in diameter and a dense rounded crown. The dull leaves with turned-under edges are borne in clusters at the end of twigs.

Flower & Fruit Date The many very fragrant, white, star-shaped, 6-mm (1/4-in) flowers are borne in clusters at the tips of twigs in the winter or spring.

Reproduction The orange-red berries maturing in the fall contain 1 to 4 round brown seeds and are sometimes so numerous that they cause the branches to bow.

Propagation Fresh clean seeds planted in a light potting soil will germinate eventually but grow slowly. Some gardeners find it does better if inoculated with some of the native soil from under the parent plant. It is difficult to transplant successfully.

Habitat & Ecological Distribution Torchwood is found in both dry and moist coastal habitats, including beach berms where some of its roots are periodically moistened by wave action. It is slow growing but persistent when established.

Uses

Ornamental The compact foliage, very fragrant flowers and attractive fruits make this shrub an underutilized candidate for landscaping.

Toxic The crushed fruits have been used to poison and stupefy fish. The shrubs of this genus have been found to contain and exude a triterpenoid called jacquinonic acid which is a strong repellent to leaf-eating ants and other insects.

Physical The very hard wood is attractive due to the conspicuous rays but is seldom used due to the small size of the tree. The inner bark has been used as a soap to wash clothes.

Ecological When planted directly on the waterfront, the dense spreading root system will help to reduce erosion from storm waves.

Notes The genus is named for the 18th-century Austrian botanist Nicholas Joseph von Jacquin who studied and named many West Indian plants. The species *arborea* refers to the treelike form of this shrub. The species *keyensis* refers to the occurrence of that species in the Florida Keys where it is called joewood. Barbasco is also used as a species and common name for this plant.

GUIANA RAPANEA, MYRSINE

Myrsine guianensis   *Myrsinaceae*

Geographic Distribution The genus of about 150 species is found in the warm regions of the world. This species occurs from South Florida through the Caribbean to Argentina.

Form A small tree to 9 m (25 ft) in height with a narrow open crown and a trunk up to 15 cm (6 in) with thin smooth grayish bark. The alternate, closely spaced, bright-green leaves cluster at the end of the stem and have turned-under edges.

Flower & Fruit Date The small greenish-white flowers with purple stripes occur singly on the twigs below the leaves in the spring.

Reproduction The many spherical 3-mm (1/8-in) blue-black fruits, each with one seed, are crowded together along the stem.

Propagation The cleaned fresh seeds may be planted in standard potting mix and with sustained patience will germinate within 6 months. Scarification of the seeds for 2 hours in concentrated sulfuric acid will accelerate germination. Young seedling plants are sometimes abundant under their parents and may be transplanted from the wild. The slow growth rate is speeded up with fertilizer.

Habitat & Ecological Distribution This tree is found in full sun and can tolerate heavy shade in moist or dry coastal forests. Tolerant of wet soil, salt, drought and light frost.

Uses

Ornamental This plant has been formed into a hedge by careful pruning. A row of these columnar trees is a good choice for shading a south wall of a house or bordering a driveway.

Medicinal The hydroxybenzoquinone derivatives embelin and rapanone found in the bark show antibiotic activity against both gram+ and gram- bacteria. They also show some potential for treating chloroquine-resistant malaria.

Physical The hard strong wood is used for general construction, crates, cabinetry, carpentry, fuel and posts. Indians have mixed the leaves with smoking tobacco as an extender.

Notes *Myrsine* is used as both the genus and common name for this plant and is derived from the Greek word for myrrh. The name *Rapanea* is also used for the genus and the common name and is derived from the native Indian name for the plant. The species name refers to its original description from the Guianas.

FRANGIPANI

Plumeria alba   *Apocynaceae*

Geographic Distribution The genus has 45 species in the tropical Americas. This species is found in the wild from Puerto Rico through the Lesser Antilles. The related *P. rubra*, a native of Mexico, has been introduced throughout the warm parts of the world.

Form A small open-crowned tree to 8 m (25 ft) tall. The 15- to 30-cm (6-to-12 in) long lanceolate leaves occur at the tips of blunt forked branches. The cultivated form has a broader leaf.

Flower & Fruit Date The clusters of showy white fragrant flowers are most abundant in the spring but may occur in any month. Cultivated *P. rubra* varieties may have white, yellow, red or purple flowers. The scent has been described as being tremendously sweet, yet fresh and suave, but tenacious beyond belief.

Reproduction The fruit consists of two long brown pods containing many winged seeds which are dispersed by the wind.

Propagation Rooting success of cuttings is very high if the cut branches are allowed to dry for several weeks before implanting their ends 3 inches deep in soft soil. The growth rate is rapid and it flowers when young. It cannot stand continuously moist soil.

Habitat & Ecological Distribution Broadly distributed from rocky sea cliffs to the upper elevations of dry islands. It is drought resistant but cold sensitive. Frangipani requires full sun and is shaded out by other plants in more moist sites.

Uses

Ornamental Many of the forms have been used as landscape plants due to the attractive flowers and pleasant perfume.

Medicinal The milky sap has been used to seal fresh wounds. A tea prepared from the leaves has been used (dangerously) as a purgative, to expel intestinal parasites and externally to treat cutaneous diseases. A leaf bath has been used to treat erysipelas and a bark bath to treat piles. Introduction of a blunt branch tip into the female reproductive tract is reported to induce abortion. The root bark has been used to treat

gonorrhea and other venereal diseases. Modern medicine has found extracts to be locally anesthetic, cardiotonic, bacteriostatic and active against the polio virus. Alcohol extracts of the petals have been found to have antibiotic action against many common pathogenic bacteria.

Toxic The milky sap is poisonous.

Physical The light brown wood is hard and tough. It has been used in carpentry in areas where the tree reaches sufficient size to produce useful lumber. The blossoms are slow to wilt and are used to produce leis and other floral decorations. The scent extract from the flowers has been widely used as a component in floral perfumes. The abundant milky sap produces a high quality latex on exposure to air.

Notes The common name is from the French and refers to coagulated milk. The generic name honors Charles Plumier, a French botanist. Individual plants may aestivate and shed all their leaves, leaving stark bare branches for a month or more. This leaf loss is commonly blamed on the large sphinx moth's velvety black caterpillars with brilliant yellow bands and a red head. Actually the leaves are usually consumed just prior to when they would be dropped. It is noteworthy that nothing eats the conspicuous caterpillars due to their acummulation of toxic compounds derived from the host tree. The physiologically active constituents of the plant include plumieride (purgative), plumericin (antibiotic), fulvoplumierin (bacteriostatic particularly of tuberculosis), cerotonic acid, calcium plumeriate, plumieric acid, ursonic acid and lupeol. The purgative glycoside plumierin has been extracted from the bark but is absent from the leaves. Also present in the bark is amyrin acetate, beta-sitosterol, and scopoletin.

PERIWINKLE

Catharanthus roseus   *Apocynaceae*

Geographic Distribution The genus has 5 species found in the tropics, especially Madagascar. Originally described from Madagascar, but believed (wrongly) to be a native of the West Indies, the periwinkle was widely introduced and naturalized in warm parts of the world by the end of the 18th century.

Form A herbaceous perennial to 60 cm (2 ft) tall with glossy opposite leaves.

Flower & Fruit Date The showy flowers, which may be white, rose or white with a deep-red throat, are present throughout the year.

Reproduction The small inconspicuous pod contains the small black seeds. Once started in an area they will reseed and maintain themselves for many years.

Propagation The seeds germinate and grow readily. Cuttings root readily and assure continuation of a favored flower color. Heavy frost kills periwinkle, requiring reseeding in the spring. In warmer climates continuous growth results in leggy plants which need to be periodically cut back. This plant generally has a low demand for nutrients and seldom shows much response to fertilizer. Germination is enhanced by up to 0.1% salt in the irrigation water. The subsequent growth continues uninhibited at 4 times that salt concentration. Salt is tolerated at much higher levels but results in reduced growth rates. When grown commercially as a source for alkaloids, a dry soil regime results in higher alkaloid content in the roots and stems, but the reduced growth rate results in an overall lower yield. A soil moisture deficit of about 50% maximizes the production of the alkaloid vincaleukoblastine.

Habitat & Ecological Distribution This plant prefers the full sun present on the open disturbed sites on which it volunteers and reseeds itself. It will grow on almost any soil and tolerates open shade, wind and salt spray.

Uses

Ornamental Widely planted in open area beds, on banks and borders of lawns. It thrives on dry seaside sites and requires little care. The horticulture industry has developed several prostrate low-growing forms which are particularly suited for use as a ground cover. It thrives as an indoor potted plant if placed in a sunny spot.

Medicinal The leaves have been used to prepare a tea as a home remedy for diabetes, diarrhea, excessive menstrual flow and high blood pressure. Tea from the flowers has been used to treat asthma. The mature leaves and roots induce vomiting and have been used as a worm treatment. An extract of the white red-eyed flower has been used to treat inflamed eyes. Of the 90 alkaloids isolated from this plant, modern medicine has discovered 6 to be useful in cancer therapy of which two are available commercially. Vincristine is used to treat leukemia in children, while vinblastine is used to treat Hodgkin's disease and choriocarcinoma. The characteristic effect of these drugs is to stop cell division at metaphase. These two drugs in combination with other drugs and periwinkle alkaloids are being used in several multi-drug cancer treatments. Two drugs prepared from periwinkle, Vinculin and Covinca, have been marketed in several countries for treatment of diabetes. They have not received approval from U.S. authorities as being safe and effective. Other alkaloids have been discovered which are hypotensive, diuretic, central nervous system depressants, and stimulants. Other extracts have been found to be active against influenza, vaccinia, and polio viruses.

Edible The very young leaves are used as a vegetable.

Toxic The mature leaves are toxic due to the presence of numerous alkaloids. Livestock have been poisoned as a result of eating the foliage. Smoking of the dry leaves may produce hallucinations due to the presence of ibogaine alkaloids.

Physical Plantations for production of pharmaceutical material have been established in Africa, India, Australia and Europe.

Notes The genus is from the Greek word meaning "pure flower." The species name refers to the rosy or pink flowers. This plant has also been called *Vinca rosea.* Many plant products of great value to humanity have simply been overlooked when investigations looking for a particular characteristic ignore a secondary useful trait. A serendipitous exception to this occurred when researchers who were investigating extracts of *Catharanthes* for the folklore claim of effectiveness in diabetic treatment discovered the associated depression of white blood cell production. This observation led to the development of Vinblastin. At the same time, other scientists who were screening drugs on mice for leukemia treatment discovered the alkaloid that was to be developed as Vincristine. The National Cancer Institute failed to find any antineoplastic activity in *Catharanthus* alkaloids because they were not using tests which detected the alkaloid's activity. Many projects to screen for certain plant traits overlook a resource due to narrow focus in a testing program. The alkaloids catharanthine, lochnerine, tetrahydroalstonine, vindoline, vindolinine dihydrochloride and leurosine sulfate are known to lower blood sugar and may have a future in treating diabetes.

OLEANDER

Nerium oleander   *Apocynaceae*

Geographic Distribution The genus has several species in the Old World. A native of the Mediterranean and Asia, the oleander has been introduced and sometimes naturalized throughout the warm parts of the world.

Form It is an upright shrub often with many straight stems to 20 feet tall. The dark-green leathery lanceolate leaves are up to 3 inches in length with conspicuous lighter veins.

Flower & Fruit Date In constantly warm climates, oleanders continuously bear single or double white, yellow, pink, red or purple flowers with 5 petals. Chilly winters restrict flowering to summer months.

Reproduction The paired creased slender pods contain many fuzzy flattened winged seeds which are disbursed by the wind.

Propagation Young but woody stems cut in the summer and treated with rooting hormone placed in moist soil root readily. Fresh seed with the fuzzy coating removed will germinate within about two weeks of planting. The red variety is most resistant to cold damage.

Habitat & Ecological Distribution Oleander has been planted and continues to thrive in many habitats. It can withstand light frost and is very resistant to damage from persistent strong winds. It is very tolerant of salty soil and prolonged drought.

Uses

Ornamental The oleander's inclination to sprout and flower vigorously after pruning encourage it to be used as borders, hedges, screens or as single focal specimens.

Medicinal Extracts of the plant contain cardiotonic heterosides with an action similar to digitalis and have been used (hazardously) in various home remedies. Externally the leaf tea has been used to treat skin problems and ectoparasites. In Africa the ground leaves are used to treat falling hair and itching. Fresh leaves have been applied to reduce tumors, and, interestingly, modern medicine has found the leafy stems to be useful in treating adenocarcinoma. Extracts of the leaves and flower petals have shown antibacterial and antiviral properties.

Toxic The sap, leaves, flowers, bark and roots are all extremely toxic due to the presence of cardiac glycosides. Cattle have been killed from consuming 12 g (1/2 oz) of leaves, and the milk from cows consuming the leaves is toxic. Human poisoning from using the straight twigs as a skewer to roast wieners has occurred repeatedly. Alexander the Great lost a number of men from his army when they camped near a watercourse in which oleander was growing. The men roasted meat on spits cut from oleander branches and many died from the resultant poisoning. The smoke from burning oleander is also toxic. Symptoms include vomiting, dilated pupils, vertigo, convulsions, respiratory difficulties and slow pulse, eventually leading to coma and death. Oleandroside and nerioside are the principle glycosides.

Physical The yellow wood is hard and brittle, but seldom reaches a useful size. The toxic properties have been utilized in making insecticides and rat poison from powdered materials and extracts.

Ecological Because of its ability to grow in a wide variety of substrates it is useful in shore stabilization and as a windbreak. The toxic but attractive flowers sometimes trap insects, particularly screwworm flies.

Notes *Nerium* is the original Greek name for oleander. The species name refers to the flowers resembling those of the genus *Olea*. The nectar from the flowers produces a toxic honey. Oleandrin, the major cardiac glycoside, is the 3 glycosido-16-acetyl derivative of gitoxigenin with the formula $C_{32}H_{18}O_9$. It is listed as cardiotonic and diuretic in the Russian Pharmacopoeia. Oleandrin in appropriately small doses regularizes cardiac flutter and is rapidly eliminated by the kidneys. The other major cardiac glycoside is nerioside. Additional cardiac glycosides present are urechitoxin, neritaloside, desacetyoleandrin and adynerin. Other compounds identified from the leaves are neriantoside, adyneroside, folineroside, oleandroside, ursolic acid, chlorogenic acid, flavanoid resins, beta-sitosterol and tannins. The bark additionally contains rosaginoside and corteneroside. Agigoside and related compounds have been identified in the seeds.

WILD ALAMANDA
Urechites lutea   *Apocynaceae*

Geographic Distribution   The genus of several species is found in the Americas from the subtropics to the equator. This species is found in southern Florida and the West Indies.

Form A slender but woody shrubby vine to 3 m (10 ft) or more long with milky sap. The opposite herbaceous oblong glossy leaves may be up to 9 cm (3 1/2 in) long.

Flower & Fruit Date The 6 cm (2 in) showy yellow flowers, which sometimes have red throats, emerge sequentially on short stalks throughout the year.

Reproduction The fruit is a curved pod, usually paired, to 20 cm (8 in) long. The seeds each have a parachute and are widely dispersed by the wind when the pods split open.

Propagation New plants can be grown from seed or cuttings.

Habitat & Ecological Distribution It is found climbing on shrubs or over rocks on beaches, rocky headlands and salt flats. It is very salt-tolerant and retains its glossy green foliage in an attractive manner even in harsh droughts.

Uses

Ornamental It is an excellent native ornamental which responds well to training on pillars or trellises. It has been generally overlooked as a decorative plant.

Toxic Consumption of the flowers or leaves produces burning of the mouth and throat, nausea, diarrhea, drowsiness, paralysis, convulsions and sometimes heart failure. The toxic component is the cardiac glycoside urechitoxin.

Notes The genus name is derived from the Greek *echites*, meaning "tailed" or "twining." The species *lutea* refers to its yellow flower color. Wild unction is another common name used for this plant. Human poisoning should be treated similarly to digitalis poisoning with electrocardiograph monitoring.

BEACH MILK VINE
Matelea maritima   *Asclepiadaceae*

Geographic Distribution   The genus of about 130 species is found in the warm parts of the New World. This species is found throughout the Caribbean.

Form   A climbing twining vine with thin stems to 3 m (10 ft). The opposite leaves are heart shaped 10 cm (4 in) long by 7 cm (3 in) wide with petioles 3 cm (1 in) long.

Flower & Fruit Date   The flowers are borne at the bases of the leaves in winter and spring.

Reproduction   The seeds have tufts of hairs to aid wind dispersal and are borne in elongate warty pods.

Propagation   The seeds germinate well if barely covered with moist potting soil.

Habitat & Ecological Distribution   This plant may be found climbing on coastal rocks and shrubs or more inland climbing to the sun on any available substrate.

Uses
   Ornamental   This plant has the potential of being an interesting addition to a natural garden in which it is allowed to climb on fences, walls or other woody plants.

Notes   The genus is from the Latin referring to the pot or vessel shape of some of the flowers in some species. It has also been known by the genus *Ibatia* The species name tells us the plant is often found near the sea.

BEACH MORNING-GLORY
Ipomoea pes-caprae   *Convovulaceae*

Geographic Distribution   This is a cosmopolitan genus of 500 species. This species is one of the world's most ubiquitous beach plants, being found worldwide on tropical shores.

Form   A sturdy much-branched trailing vine with a thick starchy root. A single vine may extend over 30 m (100 ft), frequently rooting at the nodes. The thick leathery leaves are folded on the midvein and broadly notched at the tip. The new leaves have red nectaries on the petiole at the base of the blade which are attractive to ants. The nectaries turn black and lose their attractiveness as the leaves age. The long straight runners with regular alternate leaves account for the common name railroad vine.

Flower & Fruit Date The long-stalked, bright pinkish-purple flaring morning glory-type flowers are more abundant in the summer, but may be found in any month of the year. The odorless self-incompatible flowers open early in the morning and begin producing an abundant nectar flow. They usually close in the afternoon and last only one day. Pollination is primarily by carpenter bees (*Xylocopa*) and various other *Hymenoptera, Lepidoptera* and *Coleoptera*. The pollinators are attracted by the reflecting ultraviolet patterns of the flowers which contrast with the darker leaves and stems. Ants continue to harvest the nectar after the flowers have closed.

Reproduction A single brown fuzzy seed is found in each of the chambers of the four-parted woody seed capsule. Natural reproduction seems to take place predominantly by the numerous runners vigorously sending down deep, widely dispersed roots at the leaf nodes. While seeds remain viable when floating in seawater for up to 6 months, the germination percentage of untreated seeds seems to be low. Thus, relatively few new plants are started by germination of seeds. Colonization usually proceeds from a small number of plants started by seeds followed by extensive vegetative reproduction.

Propagation The rooted nodes grow vigorously if transplanted, and stem cuttings treated with a rooting hormone start with a high rate of success. The seeds germinate and grow more successfully if scarified or soaked in water for several days before planting. The vine will thrive in many circumstances but does best when growing in full sun on sandy soil. The growth rate is fast but is further accelerated by adequate water and fertilizer.

Habitat & Ecological Distribution One of the earliest invaders of disturbed sandy areas, it may be found stretching across even the most recently deposited beaches and sand spits. It seems likely that being scuffed about by storm waves and shifting sands accomplishes scarification of the tough testa, leading to the high germination rates often seen after storms. This plant requires full sun and is shaded out as secondary plants become established on its sites.

Uses

Ornamental May be used as a ground cover, or climbing on a trellis to protect other plants from wind or salt spray. Pruning is often required to keep it under control in planned landscape situations.

Medicinal The leaf or root tea has been used as a home remedy to treat fevers, kidney problems, gout, and rheumatism. It is used as a bath to treat sores, wounds and to reduce swelling of the legs. The boiled tubers are said to be diuretic and aid diseases of the bladder. Stem extracts have been used successfully against tumors. Based on the folk use of this plant by Thai fishermen to treat jellyfish stings, experimental studies have demonstrated an antagonistic and antihistiminic effect of leaf extracts to jellyfish toxin the equivalent of Benadryl.

Edible The leaves and roots are reportedly edible but dizziness and other adverse effects result from consuming large amounts. Most herbivores eat the foliage reluctantly. Cows fed on the leaves give a tainted milk.

Toxic The toxicity does not seem to extend to all invertebrates. The bruchid beetle *Megacerus leucospilus* lays its eggs on the developing seeds. One larvae completes its life cycle in each seed and chews a large round hole to exit as an adult after about 2 months.

Physical The crushed leaves have been used in laundering.

Ecological The thick mats formed by this vine can be very effective in erosion control. The plant seems to be able to sense the high-water mark and changes direction of growth before reaching it. It is frequently found growing with *Canavalia* but is dominant to it closer to the water and on other highly disturbed sites.

Notes The genus name is from the Greek word meaning "ivy" and a similar word meaning "worm-like twining." The species name is from the Latin meaning "goat's foot," referring to the leaf shape. Bay hop is another commonly used name for this plant. A thorough chemical analysis of extracts from this plant shows the presence of pentatriacontane, glycerin, triacontane, alpha sterol, behenic acid, melissic acid, formic acid and acetic acid; the fatty acids butyric, caproic, caprylic, capric, myristic and oleic; chlorophyll, sodium and potassium chloride, and a catechol tannin. The seed oil contains stearic, oleic linoleic and linolenic fatty acids. The hallucinogenic alkaloids detected in other members of this genus seem to be represented by the indole alkaloid ergotamine, but some investigators find no presence of alkaloids.

Moon vine

Ipomoea macrantha   *Convovulaceae*

Geographic Distribution Found throughout the world on tropical shores.

Form A tough twining vine to 12 m (40 ft) long growing prostrate on sand or rocks, and often climbing on adjacent vegetation. The heart-shaped leaves are alternate.

Flower & Fruit Date The white petals form a tube 1 cm (1/3 in) in diameter and 10 cm (4 in) long which flares to form a discoidal apex on the flower

5 cm (2 in) across. The flowers open early in the evening and begin to wilt at sunrise. Flowering is intermittent through the year.

Reproduction The 4 dark-brown fuzzy seeds are borne in a dry capsule whose segments split open when mature.

Propagation The large virile seeds are the logical choice for starting new plants.

Habitat & Ecological Distribution Found along the shoreline growing on sand, among cobbles, or even on rocky coasts with pockets of soil.

Uses

Ornamental This plant has great potential as a ground cover and as a binder of drifting sand.

Medicinal It has been used as a home remedy to treat fevers, bruises and swelling of sprains. Modern medicine has found strong uterine stimulating effects from the seeds, probably due to their ergonovine (ergometrine) content. A resin extracted from the stem has been found to have a strong purgative effect.

Edible The young leaves are edible when cooked.

Toxic The seeds are toxic and may produce hallucinations and other psychological disturbances in humans.

Ecological This is one of a group of early colonizers which are very effective in invading, colonizing and stabilizing bare sand. It is usu-ally eliminated by hurricanes but rapidly re-establishes itself from rhizome fragments and seeds buried in the sand.

Notes This plant has also been known as *Calonyction tuba* and *Ipomoea tuba*.

The oil from the seed includes linoleic, palmitic, arachidonic, oleic, stearic, and linolenic fatty acids of which 37% are saturated. The foliage contains the dihydric alcohol ipuranol which is also found in olive bark. Hallucinogenic indole alkaloids found in the seeds of several members of this genus include lysergic acid amid, isolysergic acid amide, chanoclavine, elymoclavine, lysergol, ergometrine, ergometrinine, agroclavine, penniclavine, lysergic acid-a-hydroxyethylamide and the two peptide-type ergot alkaloids ergosine and ergosinine. The structures of these alkaloids are related to LSD and are known to have similar effects. The psychological disturbances produced by these chemicals were known and used by the pre-European Aztecs and Zapotecs of Mexico. Synthesis of the indole alkaloids takes place in the leaves. They are then translocated throughout the plant and accumulate in the seeds.

Pigeon berry
Bourreria succulenta   *Boraginaceae*

Geographic Distribution The genus has about 30 species in the tropical Americas. The species occurs throughout the Caribbean, but not in the Bahamas.

Form A small tree to 8 m (25 ft) tall with a trunk to 15 cm (6 in) in diameter. The alternate elliptic leaves are 5 to 12 cm (2 to 5 in) in length.

Flower & Fruit Date The 12-mm (1/2-in) white tubular flowers are born in clusters primarily in the fall but some may be found in any month.

Reproduction The 9-mm (3/8-in) diameter fruits are orange-red and fleshy when ripe containing 4 ridged seeds.

Propagation There is considerable variation in germination success with seeds from different trees, so some experimentation with several seed sources is often necessary. Cuttings are difficult to start.

Habitat & Ecological Distribution The pigeon berry is most often found in coastal and low elevation forests. It seems to require full sun to thrive and can withstand very dry sites.

Uses

Ornamental Its limited size, hardiness and abundant red berries have made this a common yard tree throughout its range.

Medicinal A tea from the inner bark is used to soothe inflamed eyes and has been used to treat inflammations of the mouth. An infusion of leaves in rum is reputed to be an aphrodisiac.

Edible The berries are edible but have little flesh. The flowers of a similar species in Mexico have been used as a flavoring of fermented beverages, preserves and confections as well as a perfume in tobacco.

Toxic The leaves are toxic to goats.

Physical The light-brown, hard, heavy wood is straight-grained, fine-textured, easily worked and finishes smoothly. Because of its small size it is most used for posts and for fuel.

Ecological The abundant flowers are attractive to bees. Many species of birds eat the berries.

Notes The genus is named in honor of J.A. Beuer, a Nuremburg apothecary. The species name refers to the juicy flesh of the fruit. The voluminous nectar produces a light mild honey. Calcium oxalate crystals are found in the wood of members of this family.

Geiger tree
Cordia sebestena   *Boraginaceae*

Geographic Distribution The genus has 250 species in the warm parts of the world with the greatest diversity in tropical America. The natural range of the geiger tree is from Florida through the West Indies to northern South America. It has been introduced throughout the tropics.

Form A small deciduous tree with a compact crown growing to 8 m (20 ft) tall with a trunk of 15 cm (6 in) and large raspy leaves.

Flower & Fruit Date The clusters of orange-red, trumpet-shaped flowers borne on the ends of branches may be found in any month.

Reproduction The smooth white oval 3 cm (1-in) fruit have a thin layer of pulp over a thick-walled stone containing one to four seeds.

Propagation The fruits may be soaked until rotten. Then the seeds should be removed and cleaned before planting in a light soil. Germination can take from several weeks to over 6 months. Various forms of scarification may shorten the germination time. The strong tendency for seedlings to damp off should be countered with limited watering, a fungicide treatment or the inclusion of some salt in the irrigation water. It is advisable to use fruit from different trees until a vigorous seed source is discovered. Cuttings are difficult but air layers may be used to start new trees. Transplant success is generally good. Once established, the growth rate is modest. It prefers a dry environment; excess watering will lead to problems. In certain areas the foliage is ravaged by beetles, but they can be controlled by a light spraying with rotenone.

Habitat & Ecological Distribution The geiger tree is salt- and wind-tolerant and survives well in dry areas with poor, alkaline or marl soils. This tree is very cold-sensitive but quite resistant to hurricane damage. It needs full sun.

Uses

Ornamental On many pieces of challenging windward ocean-front land it is one of the few trees that will thrive and flower continuously in the face of adversity. It has been used as a street tree, for borders and as a free-standing yard tree.

Medicinal Teas and syrups prepared from various parts of the tree have been used to treat coughs, dysentery, incontinence, malaria, venereal diseases, and as a tonic to sharpen the appetite.

Edible The ripe fruit is edible but fibrous and not very sweet. It is improved by cooking.

Toxic The unripe fruit is reported to be emetic.

Physical The beautiful pleasantly scented light-brown sapwood and dark-brown heartwood are hard and finely textured. The wood has been used in carpentry, turnery, small cabinetry and for musical instruments. Members of this genus frequently have silica grains embedded in the wood.

Ecological The tubular flowers have the characteristic shape of those used by hummingbirds, and indeed a geiger in flower will often be attended by a series of buzzing hummers.

Notes The genus is named after Valerius Cordius, a 16th-century German botanist. The species name is derived from the Arabic and was originally applied to a different plant. John Geiger, a seaman and wrecker, first introduced his namesake tree to Key West from the West Indies in the early 19th century. The tree he planted still survives on the grounds of what is now known as the Audubon house on Whitehead Avenue. Audubon used a flowering twig from this tree in his painting of a white-crowned pigeon. The hardy and attractive *Cordia rickseckeri* is native to Puerto Rico and the Virgin Islands.

SEASIDE HELIOTROPE

Heliotropium curassavicum   *Boraginaceae*

Geographic Distribution The genus has 200 species found in the warm parts of the world. This species is native to the southern U.S., Central America and the West Indies and it has been introduced and naturalized in the Mediterranean basin, Australia and India.

Form A branching, sometimes prostrate, rubbery perennial herb to 60 cm (2 ft) long. The alternate leaves are thick and juicy. When grown in a hot dry environment where water is only available in the soil, the growth form has a reduced leaf area and is taller with thinner, more steeply inclined leaves.

Flower & Fruit Date The small white flowers are borne year-round in scorpioid clusters.

Reproduction The seeds are surrounded by a corky coating which allows them to float and be widely dispersed by water.

Propagation Both cuttings and seeds can be used to start new plants. While they can tolerate almost any soil, a rich, slightly acid soil promotes lush growth.

Habitat & Ecological Distribution This plant grows in sandy soils on or near beaches, salt flats and as an understory plant in mangroves when openings in the canopy allow sufficient light. It is able to withstand the heat, low humidity and saline soils of Death Valley wherever soil moisture is present.

Uses

Ornamental It is grown as an ornamental in Hawaii, and the lush blue-green foliage makes it a candidate for a ground cover on damp salty soils. *Medicinal* The leaf tea is used as a home remedy for coughs, asthma, gout, rheumatism, arteriosclerosis and scorpion sting. Indians have used the powdered dried root to treat wounds and sores. Extracts from similar species have been found to lower blood pressure in animals without central nervous system effects.

Toxic Many alkaloids have been dis-covered in this plant. The pyrrolizidine alkaloids have caused mortalities in cattle and horses which grazed on this plant. Goats may survive the consumption of small amounts of this plant but the toxins are excreted in their milk. It may become so toxic that even locusts refuse to eat it. The esters of the aminoalcohol heliotridine are known to produce liver cancer.

Notes The genus is from the Greek word meaning "turning to the sun." The species refers to the island of Curacao from which the plant was first de-scribed. The greatest alkaloid content of the plant is at the beginning of flowering with the highest concentrations in the roots and flowers. The pyrrolizidine alkaloids present are extremely variable from one location to another but most are trachelanthamidine esters (curassavine, coromandaline, heliovicine, supinidine and retronecine), heliotrine, heliotridine and indicine. Minor alkaloids include heliocurassavine, heliocoromandaline, heliocurassavicine, helio-curassavinine, curassavinine, coromandalinine, heliovinine and curassanecine. The plant also contains beta-sitosterol and the lactone hydroxy hexacosanoic acid. The seed oils are composed of palmitic, arachidic, oleic, linoleic and behenic acids.

Bay lavender
Argusia gnaphalodes   *Boraginaceae*

Geographic Distribution Depending on how the nomenclature is arranged this may be the only species in the genus which is found in Florida, Mexico and the Caribbean; or a member of a genus of over 100 species found worldwide in the tropics. This species is found from Bermuda south through the Caribbean.

Form A small shrub to 2 m (6 ft), but usually less than a meter in height with fleshy branches. The slender, soft, downy gray leaves form rosettes at twig tips.

Flower & Fruit Date The small white bell-shaped flowers occur in clusters on branch tips year-round.

Reproduction The seeds are surrounded by a corky head which allows wide dispersal by water.

Propagation Branches which touch the soil take root by natural ground layering and when detached are easily transplanted. Intentional ground layering, in the native soil or in strategically located pots, works well. Transplanting of seedlings requires that a larger than normal root ball be taken with the plant, but uprooting seedlings is becoming increasingly restricted by law in many locations. Six-inch cuttings treated with auxin B rooting hormone and planted 12 cm (5 in) deep should be placed in heavy shade and watered daily. Another method for rooting cuttings is to treat them with 8000 ppm IBA and maintain them in a mist house while rooting in a 3:1:1 mixture of sphagnum, perlite and vermiculite. The shade should be gradually reduced to full sun by 3 months and the watering reduced to once every 3 days. When the plants are over 30 cm (12 in) tall they may be planted out on a dune. Transplanting should be 3 cm (1 in) deeper than the soil in the pot. Watering once or twice a week if there is no rain will help them become established. A light application of a balanced fertilizer once or twice a year improves the growth rate on sterile dune sands. Growing plants from seed is difficult because the germination rate is very low. A bit of salt in the media helps prevent root rot and damp off.

Habitat & Ecological Distribution This salt- and drought-tolerant plant is found on exposed sand or cobble beaches and rocky coasts just above the wave wash zone.

Uses

Ornamental This interesting plant which needs little care has become increasingly scarce in southeast Florida. It is now available from native plant nurseries and is an ideal border plant for beach landscapes.

Medicinal A tea prepared from the leafy twigs is taken to treat fevers, gonorrhea, syphilis, bladder stones, kidney problems, fish poisoning, rheumatism, and to induce abortions.

Physical The smoldering leaves are reputed to drive fleas and other vermin from a house.

Ecological The inclination to grow in spreading clumps makes this a good sand binder.

Notes The species name refers to the resemblance of this plant to those in the genus *Gnaphalium*. The generic names *Mallotonia* and *Tournefortia* have also been used for this plant. It is also called sea lavender.

BLACK MANGROVE
Avicennia germinans   *Verbenaceae*

Geographic Distribution The genus of about 10 species is found worldwide in the tropics. The black mangrove is found on both coasts of tropical America and west Africa.

Form A small tree or shrub to 36 m (120 ft) high with a rounded crown and trunk to 1.8 m (6 ft) but usually much smaller. The lance-shaped leaves

with a smooth yellow-green upper surface are gray-green on the bottom. The many lateral roots commonly send up fleshy pencil-sized pneumatophores with abundant lenticils. The pneumatophores commonly extend 6 inches or more above the saturated soil. The height increases with average water depth and allows the submerged roots to obtain atmospheric oxygen.

Flower & Fruit Date The small white flowers with yellow throats are borne in clusters near the branch tips throughout the year. They are fragrant and pollinated by insects.

Reproduction The fleshy green seed capsule germinates on the tree before

falling off. Within a few days the pericarp is shed and the cotylydens expand. It may then take root or be carried to another location by tidal currents. It can survive floating in seawater for over 4 months and considerably longer in fresh water. While the adult plant can tolerate hypersaline conditions, the seedlings require salinities of seawater or less and must be exposed at low tide for early development. Water temperatures over 40°C (104°F) are lethal to young seedlings.

Propagation The seeds sprout and grow readily when placed on edge in moist soft media. Pots may be placed in a shallow pan of water to insure the required continuous moisture. Transplant is possible with desirable root balls being 1/2 the diameter of the tree height. Vigorous pruning reduces post-transplant mortality and speeds recovery. The tree sends up strong new sprouts from cut stumps. Both transplanted trees and newly planted seedlings benefit from a high-nitrogen slow-release fertilizer and a mulch of seagrass. Under good growing conditions young trees may add 60 cm (2 ft) in height per year.

Habitat & Ecological Distribution Found primarily in salty, silty, saturated soils along tidal coasts. This is the most cold-tolerant of the mangroves and can grow in soil with a salinity of 100 parts per thousand. It is usually the dominant species where soil salinities exceed 40 parts per thousand. The black mangrove is found on higher drier soils than the red or white mangrove, and it is killed if impoundments or flooding submerge the pneumatophores for a long period of time. It is the most frost-tolerant of the mangroves, but is killed by freezing temperatures. It requires full sun. Because of its shallow lateral roots, entire stands may be tipped over by hurricanes.

Uses

Ornamental The black mangrove has been planted as a hedge and as a privacy screen in low salty damp spots. It is most suited to the waterfront of quiet canals and lagoons.

Medicinal A tea prepared from the bark has been used to treat ulcers, hemorrhoids, diarrhea and tumors.

Edible The cooked sprouted seeds are reported to be edible. The seeds contain 84% carbohydrate and 7% protein.

133

Toxic The raw fruits are toxic but have been used as a famine food after careful cooking treatment to remove the toxins. The wood contains an oil which irritates the skin of woodworkers.

Physical The dark-brown wood is hard, heavy and strong with a coarse grain. It is resistant to decay but susceptible to termites. It has been used for beams and doorframes, furniture, gunstocks, in marine construction, for utility poles and crossties and as fuel. It is suitable for paper pulp but must be mixed with other woods due to its short fiber length. The bark, containing 6% to 13% tannin, is used in tanning and as a source of a red dye.

Ecological This tree is an important soil stabilizer, promoting deposition of sediments. The leaves falling from stands of these trees contribute to the productivity of adjacent marine ecosystems that support commercially important species of fish and crustaceans. The dense canopy developed in stands of black mangroves provides roosting and nesting habitat for many species of birds.

Notes The genus is named for the 10th-century Persian physician Avicenna. The species *germinans* refers to the characteristic mangrove germination of the seeds before they fall from the tree. The black mangrove has also been known as *Avicennia nitida.* The abundant nectar is vigorously sought by bees and produces a high-quality clear, almost colorless honey. Excessive salt in the sap may be secreted in a saturated solution through microscopic glands on the undersides of the leaves. Under humid conditions the underside of the leaves is sticky with salt, but when the moisture dries out the wind blows the salt away. The heartwood contains a lapachol compound that gives it a yellowish color.

HAGGARBUSH
Clerodendrum aculeatum   *Verbenaceae*

Geographic Distribution The genus of 400 species is found in the warm parts of the world with the greatest diversity in Asia and Africa. This species is found in the Caribbean, Bahamas, Bermuda, and has been introduced to Hawaii.

Form This is a plant of extremely variable growth form. Usually a viny shrub with widely spreading branches, sometimes climbing over other trees and shrubs, it may form a tree 6 m (20 ft) high with a trunk 15 cm (6 in) in diameter. Two or three short curved spines occur below the opposite 1 to 5 cm (1/2 to 2 in) long elliptic leaves at each node.

Flower & Fruit Date The clusters of small white flowers with long stamens may be found at any time of year, probably depending on rainfall.

Reproduction The shiny black 6 mm (1/4 in) fruits are on peduncles to 2.5 cm (1 in), fleshy with 4 grooves, and contain four brown seeds.

Propagation This shrub seems to be seldom propagated but can be grown from seeds.

Habitat & Ecological Distribution The haggarbush may be found in both moist and dry coastal habitats, from sandy shores to rocky headlands with minimal soil.

Uses

Ornamental It has been grown as a hedge plant and particularly lends itself to being espaliered on walls and fences.

Medicinal Leaf poultices have been used to treat skin problems and other ailments. Experimentally, an application of an extract of the leaves prevents the virus infection of tobacco and lesion-forming viruses of several other plants.

Physical The wood is hard and heavy but easily worked. It has a fine texture. A hedge of this plant strongly discourages foot traffic.

Ecological The salt and drought tolerance of the haggarbush combined with the inclination to form dense thickets make it an ideal candidate for stabilizing sand dunes. The dense thorny thickets make an ideal nesting refuge for many bird species and when near water a favorite resting place for basking iguanas.

Notes The genus is from the Greek words for "destiny" and "tree." The species name accentuates the obvious by telling us the plant is armed with prickles. This plant is also called *Volkameria aculeata.* Other commonly used English names are crab prickle, crab bush and prickly myrtle. The flea beetle *Alagoasa bicolor* feeds on this plant.

Wild Sage, Lantana
Lantana involucrata *Verbenaceae*

Geographic Distribution The genus of 150 species is found mostly in tropical America with a few species in Asia and Africa. This species is found throughout tropical America and the West Indies.

Form A herbaceous shrub to 3 m (10 ft) tall with slender twigs and toothed, rough-textured aromatic leaves. The twigs and leaves are covered with short white hairs.

Flower & Fruit Date The pink to purple flowers in flat-topped clusters about 3 cm (1 in) across are produced year-round in response to rain.

Reproduction The spherical dark-blue to purple fruits, each with a single seed, are borne in dense clusters on branch tips.

Propagation Lantana is unfortunately easy to cultivate from seed and often becomes a weed in agricultural areas. Cuttings also root readily. Young plants are easily transplanted if soil is moved with most of the roots intact. An excess of water tends to bring on a lethal fungus of the roots.

Habitat & Ecological Distribution Lantana may be found growing on sand dunes, uplands near mangroves, and in the interior on almost any type of soil.

Uses

Ornamental Several introduced species of lantana are planted as ornamentals and have escaped and become naturalized, and several others are considered to be agricultural pests. This species can be pruned into a hedge. Others are prostrate and make good ground covers. There are varieties with yellow, orange-red, tangerine, red, lavender and cream flowers.

Medicinal Folk medicine has used a tea prepared from the flowers to treat high blood pressure. The leaf tea has been used to treat fevers, venereal disease, and as a bath to relieve the itching of chickenpox and measles. The leaves pounded in water have been used as an emetic. It also reportedly induces sleep, acts as a diuretic, and regularizes menstruation.

Edible The ripe fruits have been reported as edible but should be approached with extreme caution.

Toxic The green fruits are poisonous to humans. Within a few hours of consumption of the berries, vomiting and diarrhea is produced which may lead in extreme cases to lethargy, cyanosis, depressed nervous responses, and even death. Known consumption of the berries should be treated with gastric lavage and other supportive measures until the signs of toxicity have declined. Birds and livestock can eat small amounts of the fruit with impunity. Consumption of the leaves produces photosensitivity in livestock, with lesions appearing on exposed skin with subsequent sun exposure. Continued consumption of the leaves produces chronic or acute sickness. The leaves contain the cardioactive steroid lancamarone, which is a fish poison at dilutions of up to 1 ppm. Other members of the genus have been found to contain compounds toxic to mosquito larvae. Some people show a skin irritation upon contact with the plant.

Physical The leaves and stems are used to bait fish traps and dried as kindling for starting fires. With various processing techniques the stalks yield pulp suitable for wrapping paper, straw board, writing paper and printing paper. A volatile oil with a pleasant and persistent odor suitable for use in the perfume industry has been extracted from the leaves of *lantana.*

Ecological This plant is widely dispersed by the undigested seeds remaining after the consumption of the fruit by birds and livestock.

Notes The genus is derived from the superficial resemblance of this plant to *Viburnum lantana.* The species name is derived from the botanical term involucre which describes the whorled flower bracts. *Lantana* is often used as a common name for many of the species in this genus. The leaves and bark contain lantanine which has an action similar to quinine and is reputed to have antipyretic and antispasmodic properties. One of the toxic principals in the foliage produces acute symptoms resembling atropine poisoning. The polycyclic triterpenoid lantadene, also called rehmannic acid, produces its toxic effects by blocking the excretion of bile pigments in the liver.

PINK CEDAR
Tabebuia heterophylla   *Bignoniaceae*

Geographic Distribution This genus of trees has about 100 species found in the tropical Americas. Pink cedar is found in the wild from the Greater Antilles to Barbados and has been planted as an ornamental in Florida and the western Caribbean.

Form The pink cedar forms a medium-sized tree with a narrow crown usually to 18 m (60 ft) tall with a 60 cm (2 ft) diameter but sometimes reaching 27 m (90 ft) with a trunk 1 m (40 in) in diameter. The palmately compound leaves usually have 5 long oval leaflets each up to 7 cm (3 in) long.

Flower & Fruit Date The masses of pink tubular five-lobed flowers occur primarily in the spring before the new year's growth of leaves but may be

137

found in any month.

Reproduction The dark-brown pods 3 to 8 inches long split open at maturity to release the white winged seeds for dispersal by the wind.

Propagation Fresh seeds should be planted very shallowly. Cuttings will take root but are sensitive to a necrotic fungus of the leaves. When the soil is moist, the young seedlings may easily be pulled up and transplanted.

Habitat & Ecological Distribution This tree is broadly adaptable to a wide variety of poor and degraded soils. It may make up part of the forest canopy or may be found as a single dominant tree in more harsh sites. It has an extensive and highly effective root system which may deprive adjoining plants of water under dry conditions.

Uses

Ornamental The dark-green palmate leaves and columnar canopy combined with the seasonally abundant showy flowers have made this a popular landscape tree.

Medicinal The leaf tea has been used to relieve toothache, backache, gonorrhea, fish poisoning, and as a diuretic. With the addition of several other plants it has been used as an aphrodisiac. The bark has been brewed into a tonic and used to treat bed wetting in children. The alkaloid lapachol extracted from this species has been shown to have antibiotic effects against gram+ and acid fast bacteria and fungi. It is also being investigated as a cancer chemotherapeutic agent. A closely related species has been found to contain central nervous system stimulants.

Edible The leaf tea is sometimes consumed as a refreshing beverage.

Physical The wood is strong, with a light golden color and an attractive figure similar to white ash. It has a specific gravity of .42 when dry and exceeds white ash and white oak in strength, toughness and hardness. Its working properties are excellent in response to sawing, boring, turning and veneer slicing. It is easy to glue, takes stains very well and takes a high polish. Unpainted wood develops considerable checking and is susceptible to termite damage when exposed to the weather. It has been used for handles for sporting goods and agricultural tools, furniture, cabinetwork, flooring, interior trim, boat building, paddles, ox yokes, carts, and as a face veneer in plywood.

Ecological This is probably one of the climax trees in a dry scrub forest.

Notes The genus is a Brazilian Indian name for the tree. The species name refers to the variation of leaves on the tree. *Roble*, the Spanish name for oak, is also used for this tree. This tree has several adaptations to drought which include stomatal closure at dawn, cessation of growth when water is scarce and a root system which has been experimentally shown to be extremely effective at gathering water from soil. Some trees are deformed by witches-broom disease and others may be defoliated by leafhoppers. Horticulturists and others are urged to select resistant strains for cultivation. This tree is classed as a honey plant.

W<small>HITE-ALLING</small>

Bontia daphnoides   *Myoporaceae*

Geographic Distribution This is the only New World species in this genus of an Old World family composed of 5 genera and 100 species present primarily in Australia. The white-alling is found rarely in Florida, but commonly throughout the Bahamas and Caribbean islands.

Form A shrub or small tree to 8 m (25 ft) in height with a 15-cm (6-in) trunk. The lanceolate alternate leaves and numerous ascending branches produce a dense foliage.

Flower & Fruit Date The yellow flowers with purple markings are borne in any month singly on the stem at the bases of leaves.

Reproduction The yellow fruits, shaped like pointed olives, have a thin fleshy coating over a hard stone which encloses several small seeds.

Propagation Several nurseries growing this plant from seed are finding an increasing demand. It may also be grown from cuttings. It may be grafted to the vegetatively indistinguishable genus *Myoporum*, which fills an equivalent ecological niche in the Indian and Pacific Oceans.

Habitat & Ecological Distribution The white-alling is most commonly found along rocky shorelines and salt flats, often interspersed with mangroves and sea grape. It has been widely planted and naturalized, even to 5000 feet in the Andes. It can grow and thrive on bedrock shorelines with soil only present in pockets and cracks.

Uses

Ornamental This plant has been grown as a street tree, an ornamental foundation planting, in hedges, and as a windbreak.

Medicinal The leaf tea has been used to aid labor and childbirth, kidney complaints, high blood pressure, coughs, colds and fish poisoning. It is an interesting coincidence that this plant contains mannitol, which has recently been found to be an injectable remedy for acute fish poisoning. Several different recipes combine this with other plants in making aphrodisiac teas. Externally it has been used as a bath to soak swollen feet and soothe irritated skin.

Edible A refreshing tea has been prepared from the leaves.

Toxic Leaves have been scattered in chicken houses to repel lice and other vermin, and an alcoholic extract from the leaves has been used to repel mosquitoes. Livestock avoid browsing on the foliage, which is known to contain a sesquiterpene furan called epingaione, known to cause liver lesions.

Physical The wood is hard, heavy and fine-textured but seldom reaches a size suitable for lumber. It carves easily and finishes very smoothly.

Ecological As a very salt-resistant small tree it has considerable potential for use in shoreline stabilization.

Notes The genus is named for Peter Bontius, a 17th-century Dutch naturalist. The species name relates to the resemblance to some species of *Daphne*. Other common names for this plant frequently include the word olive, such as wild olive, Barbados olive, and sea olive. The foliage contains several volatile oils, resins, cyanogenic glycosides and flavonoids.

GOLDEN CREEPER

Ernodea littoralis   *Rubiaceae*

Geographic Distribution The 7 species in the genus are found in Florida and the Caribbean. Golden creeper is found from Florida throughout the West Indies to South America.

Form A small shrub to 1 m (3 ft) tall erect or prostrate with narrow but thick glossy leaves 2 cm (1 in) in length, often turning yellow.

Flower & Fruit Date The small white to yellow tubular flowers occur year-round in the leaf axils.

Reproduction The yellow berrylike fruit is borne along the stem.

Propagation Small plants or rooted stems can be transplanted if considerable soil is taken along with the roots. Cuttings at least 12 cm (5 in) long of new growth from branch tips should be treated with rooting hormone and set out in continuously moist potting soil in open shade. Over the following 4 months the sun exposure should be increased and the watering decreased. The new plants are generally sufficiently rooted for setting out 6 to 8 months after initial potting. Ground layering branches of established plants works well in producing new stock for transplant. The branches selected for layering can be covered with native soil or by setting nodes into pots of soil placed in strategic locations. Fresh clean seeds will grow in a slightly salty sandy soil, but may take up to 5 months to germinate. The seedlings should be watered sparingly and fertilized lightly.

Habitat & Ecological Distribution This plant is almost ubiquitous; it can be found on coastal dunes, in small pockets of sandy soil on rocky shorelines and on disturbed sites inland. It thrives on poor soil, tolerates wind, salt and drought but is sensitive to frost.

Uses

Ornamental As an ornamental it should be neglected most of the time when other plants are being watered and fertilized. It can be pruned to shape as thick mounds for a seaside ground cover on open areas or banks and slopes. It is also used as a border plant.

Medicinal A tea made from the leafy branch tips is used as a home remedy for coughs.

Ecological Its roots stabilize loose sand and in small areas it serves as a windbreak and reduces the wind movement of sand. Birds and small mammals eat the fruits.

Notes The genus is from a Greek word meaning "like a young sprout," or "branched." The species name *littoralis* from Latin makes reference to its being found on seashores. This plant is also called common *ernodea*, beach *ernodea* and cough bush.

MORINDA

Morinda citrifolia   *Rubiaceae*

Geographic Distribution The genus has about 8 species primarily in the Old World tropics. Morinda is a native of India, the East Indies, and northern Australia and is now widely introduced and naturalized in the West Indies and Florida Keys.

Form A small tree to 9 m (30 ft) high with a trunk of 20 cm (8 in) and an irregular crown. The large 15- to 30-cm (6- to 12-in) sturdy dark-green leaves are in clusters at branch tips.

Flower & Fruit Date The small white flowers grow from a light-green base in all months of the year.

Reproduction The fruits of many flowers combine to form a single fleshy structure resembling a malformed pineapple containing two seeds per section. When ripe, the fruit becomes soft, foetid and translucent with a waxy surface. The seeds contain a distinct air chamber which keeps them afloat for long periods. They retain their viability after long ocean voyages.

Propagation The seeds may be planted immediately upon removal from ripe fruit.

Habitat & Ecological Distribution The morinda grows best on moist sandy soils. It is salt-tolerant and is most commonly naturalized on beach berms and the low areas behind berms.

Uses

Ornamental Its use as an ornamental has contributed to its wide distribution.

Medicinal The application of fresh, bruised, heated or ground leaves to the skin is reputed to relieve many different aches and pains. As an example, the leaves may be wrapped around rheumatic or arthritic joints or applied to external ulcers. In Cuba it is used to regulate menstruation and cure impotency. In the Caicos the root is taken with salt to dispel indigestion, while the stems are boiled and the liquid gargled for sore throats. The fruit has been used to treat diabetes, swollen spleen, liver and kidney disease, beriberi and menstrual problems. A theory explaining the wide use of this plant in medicinal applications proposes that the active alkaloid is only released from its precursors when the juice is taken on an empty stomach. Extracts of the ripe fruit have been found to show antibiotic effects in laboratory tests

142

against *Micrococcus pyogenes* and the enteric pathogens *Salmonella typhosa, S. montevideo, S. schottmuelleri* and *Shigella paradys.*

Edible The ripe fruits are juicy and edible but have a revolting acrid odor. They are mostly eaten in times of famine and by people stranded on tropical shores. The addition of sugar is said to make the fruit somewhat more palatable. The immature fruits have been cooked in curries, and the young leaves (called *mengkuda*) are eaten raw or cooked as a vegetable in Malaya. Pigs, goats and donkeys eat the ripe fruit and it is used to bait fish traps.

Physical The wood is streaked or variegated with a disinctive orange color and a greenish hue. It is hard and heavy with a fine texture and readily worked. A yellow dye may be extracted from the bark of the trunk. The bark of roots yields a red dye and the wood a yellow dye. Red, yellow, blue, purple, chocolate and orange dyes may be obtained from the young roots depending on the mordant used and the textile being dyed. Roots from mature trees have little of the dye material.

Ecological Many species of wildlife eat the flesh or seeds of the fruit.

Notes The genus is from the Latin *morus indica* meaning "Indian mulberry." The species name refers to the superficial resemblance of the leaves to those of various citrus trees. Other names used for this tree include painkiller, Indian mulberry and wild pineapple. The fruit of this tree is the specific host for the fruit fly *Drosophila sechellia.* The roots contain tannin and calcium oxalate. At least nine different anthraquinones are known from this plant including commercially important alizarin. The most abundant of these, morindone (1,5,6 Trihydroxy-2-methylanthraquinone), is found only in this plant. The root bark has a glucoside, morindine ($C_{27}H_{10}O_{15}$). The fruit contains an unusual volatile oil with 90% n-caproic and n-caprylic acids. It also has paraffin, asperuloside, ethyl alcohol and caproic and caprilic fatty acids. The fatty acid components probably account for the unpleasant odor of the ripe fruit.

SANDFLY BUSH

Rhachicallis americana   *Rubiaceae*

Geographic Distribution The single species in the genus is found in Mexico and the West Indies.

Form A woody to prostrate shrub to 1.2 m (4 ft) tall. The tiny fleshy leaves are born in tufts on woolly twigs.

Flower & Fruit Date The minute inconspicuous flowers are borne in the leave axils in all months of the year.

Reproduction The small, angular, pitted seeds occur in capsules which split open upon ripening.

Propagation Cuttings do well in moist salty sand.

Habitat & Ecological Distribution Growing immediately next to the sea in clefts of the rocky shoreline, on sandy beaches and gravelly berms, this woody little plant is very resistant to strong buffeting winds, salty soil and occasional storm wave inundation.

Uses

Ornamental In its natural habitat it assumes a great variety of forms and is certainly overlooked as an ideal Bonsai candidate.

Toxic A smoldering fire of this plant is reported to be repellent to sandflies.

Notes The genus is from the Greek word meaning "beauty of rocky shores." The species *americana* refers to its American origin. This plant is also known as wild thyme.

STRUMPFIA

Strumpfia maritima   *Rubiaceae*

Geographic Distribution This single species genus is found from the Lower Florida Keys through the Bahamas and the Greater and Lesser Antilles.

Form A bushy shrub to 2.5 m (8 ft) but usually less than 60 cm (2 ft). The very rough stems have needlelike leaves woolly underneath in small clusters.

Flower & Fruit Date The pink or white 4-mm (3/16-in) flowers with five linear petals grow in axillary clusters.

Reproduction The red or white spherical 6-mm (1/4-in) fruit contains one or two seeds.

Propagation This plant is not often artificially cultivated but it will grow slowly from cuttings.

Habitat & Ecological Distribution On rocky shorelines, sand dunes, open saline areas and inland on sunny but damp sites (often with bedrock at the surface which inhibits the growth of taller competitors).

Uses

Ornamental This small but striking plant has been overlooked by growers and landscape designers but now has the attention of a few horticulturists. It makes an intriguing specimen plant and is an excellent choice as a rock garden showpiece.

Medicinal A weak tea made from this plant is used to treat colds, stomachache and to induce passage of kidney stones. A stronger leaf tea is used to induce an abortion which leaves the woman sterile.

Toxic The smoke from this plant on a smoldering fire is said to repel mosquitoes.

Notes The genus is named to honor Karl Strumpf, an 18th-century professor. The species is named to indicate the plant's affinity for the seashore. This little bush is also called *lirio* and false rosemary. A

phytochemical screening has shown the presence of sterols, tannins, polyphenols, and flavonoid glycosides. An aqueous extract of the flowering tops given to female rats on a daily basis has significantly reduced reproductive rate. The extract of the flowering tops contained the flavanol glycoside narcissen, the first reported flavonoid from *Rubiaceae*.

SEVEN YEAR APPLE
Casasia clusiifolia   Rubiaceae

Geographic Distribution The genus has 10 species in Florida and the Caribbean. This species is found from Bermuda south to Florida, the Bahamas and the West Indies.

Form An open canopied branching shrub to 3.7 m (12 ft) tall with opposite glossy leathery leaves clustered at branch tips.

Flower & Fruit Date The fragrant star-shaped flowers with white pointed petals are borne in groups on stalks mostly in the spring.

Reproduction The fruit is a berry about 5 cm (2 in) long with numerous angled, flattened seeds. The fruits may hang on the tree for more than a year as they change from green to yellow and eventually wrinkled black as they ripen.

Propagation No matter when they are planted, most of the seeds tend to germinate in the spring. Seeds planted in the spring show up to a 75% germination rate within a month but are slow growing. Young plants may be transplanted from the field but should have most of their leaves removed to reduce transpiration until new roots are established. Cuttings and air layers should be attempted if material is available. The established plant responds strongly to supplemental fertilizer.

Habitat & Ecological Distribution Tolerant of wind, salt spray and drought, it may be found growing on rocky coasts or inland as a component of dry forest. It does best on a well-drained soil in a location with full sun. This plant does not tolerate freezing temperatures or wet soil. Germination of the seeds in the droppings of ground iguanas (*Cyclura*) is a major source of dispersal.

Uses

Ornamental One of the best salt-tolerant small trees for rocky shoreline landscaping, beachfront or any very salty dry site.

Edible The fruit is dark, soft and wrinkled when ripe. It both looks and tastes like a prune.

Physical The hard, yellow-brown, fine-textured wood is used for tool handles because of its strength.

Ecological This tree can survive storm damage and occasional seawater inundation. It is susceptible to a fungus which produces lesions on the stem and spots on the leaves and fruit. The fruit is

consumed by birds, deer, iguanas and many other species of wildlife.

Notes The name *Casasia* honors Luis de las Casas, an 18th-century Captain General of Cuba. The species refers to the resemblance of the leaves to those of the genus *Clusia*.

INKBERRY
Scaevola plumieri   Goodeniaceae

Geographic Distribution The genus has 90 species found primarily in Asia and Australia. This black fruited species is found naturally in Florida, the West Indies and South America. A white fruited species, *S. sericea*, has been introduced from the Pacific and is becoming increasingly widespread and naturalized on Gulf and Atlantic shores.

Form An erect to trailing and spreading herbaceous shrub to 1.2 m (4 ft) tall often forming dense clumps. The alternate, glossy, green, thick leaves are clustered near the branch tips.

Flower & Fruit Date The small asymmetrical fanlike whitish flowers with a yellow throat are borne in clusters among the terminal leaves. The flowers have a special structure opposite the petals to insure pollination by visiting bees and wasps.

Reproduction The glossy black juicy fruits are about 12 mm (1/2 in) in diameter by 25 mm (1 in) long enclosing a stone containing two seeds.

Propagation Transplanting stems that have naturally rooted works well. Directly planted seed will take up to 6 weeks to germinate. Vegetative cuttings have also been used for propagation. In all cases planting should be in a sparingly watered light potting soil as damp-off is often a problem. It is slow growing but hardy. Do not water large established plants.

Habitat & Ecological Distribution A dry salty sandy habitat is preferred but it will grow on a wide variety of substrates from clay to cobbles.

Uses

Ornamental The inkberry responds well to pruning and can form many interesting shapes as a single shrub or continuous ground cover. Increasingly the western Pacific form with white berries is being supplied by nurseries. It is commonly being used as a low maintenance decorative

147

planting around large buildings.

Medicinal If more than one of the bitter fruits are consumed they are likely to have a purgative and emetic effect. The roots have been used in home remedies.

Edible The young leaves are eaten as a pot herb in India.

Physical The soft snow-white pith of large plants is sometimes cut into thin flakes by the people of Southeast Asia to make artificial butterflies, flowers and other art objects.

Ecological The dense clusters formed by this plant make excellent cover for many birds and small mam-

mals. The introduced Australian species (*S. sericea*, also called *S. frutescens*) with white berries frequently used in the landscape industry seems to be supplanting the native form with black berries on many wild beaches.

Notes The genus is named after a 6th century B.C. Roman, Mucius Scaevola. The species name is derived from the Latin word for feather; it has also been credited with honoring Charles Plumier, a 17th-century French monk who explored and wrote on plants of the Americas. This plant is also known as boboron and beach naupaka. Pollination is primarily by bees and wasps of the families *Apidae, Sphegidae* and *Vespidae*. As with many plants growing on nutrient-poor soils, the mineral nutrients from older leaves are mobilized and removed to other parts of the plant before the leaf is shed. In this salt tolerant plant, the sodium, calcium and magnesium which are in surplus in the environment remain in the old leaves. Chemical analysis has detected a great variety of flavonoids in the foliage.

SEA OX-EYE

Borrichia arborescens   *Compositae*

Geographic Distribution The genus has 2 species. *Borrichia arborescens* is found on the Gulf coast and throughout the West Indies. The similar *B. frutescens* is found from the Gulf coast north to Bermuda.

Form A stiff branching compact shrub often forming dense colonies to 1 m (3 ft) tall with fleshy lanceolate leaves that may be dark shiny green or with fine silvery hairs which give them a grayish color. The leaves of the similar sea daisy, *B. frutescens*, have silvery hairs, and the tips of its flower bracts turn downward with spiny points. In the cooler parts of its range, *B.*

frutescens is an annual. The two species hybridize in South Florida.

Flower & Fruit Date The yellow flower heads about 2 cm (1 in) across occur in all months.

Reproduction At maturity the flower heads turn down and disperse the abundant seeds broadly. Dense cloned clumps result from the rapid lateral spread of the plant by numerous rhizomes.

Propagation Seeds have a viability astonishingly close to 100% and may be started in pots or planted directly in beds. Cuttings and rooted rhizomes are also effective in starting new plants. Transplant success of mature individuals is poor.

Habitat & Ecological Distribution Highly salt- and drought-tolerant, the sea ox-eye grows well in seawater-moistened sand, in salt marshes and on the borders of mangroves. It tolerates coastal temperatures throughout Florida.

Uses

Ornamental Both species can make an ornamental ground cover or be allowed to form dense rounded clumps. B. *arborescens* can be sheared into a hedge. Ecologically and aesthetically it does well as an intermediate between sea oats and salt marsh cord grass.

Medicinal A tea prepared from leaves and branch tips is used as a home remedy for colds, coughs, back pains, asthma, malaria, and fish poisoning.

Edible The leaves have been eaten in a salad with vinegar to prevent scurvy.

Ecological The seeds are eaten by many birds and rodents. Dense stands help stabilize the soil from wind and water erosion, and make excellent cover for wildlife.

Notes The genus is named for a medieval Danish botanist, Ole Borch. The species *arborescens* refers to its sometimes small but treelike form. The species *frutescens* refers to a shrubby growth habit. This plant is also known as seaside marigold, bay marigold, sea daisy and seabush.

149

TRAILING WEDELIA
Wedelia trilobata   *Compositae*

Geographic Distribution The genus of 55 species is found worldwide in warm areas but mostly in the Americas. Originally a native of America, wedelia has now been introduced and naturalized throughout the warm parts of the world.

Form A prostrate creeping herbaceous perennial with limbs up to 1.2 m (4 ft) long, but seldom over 30 cm (1 ft) tall unless climbing on some support. Frequently rooting at the nodes, it forms a dense mat. The opposite, hairy, slightly fleshy leaves are lightly toothed.

Flower & Fruit Date The bright yellow 3-cm (1-in) wide flower heads with darker yellow centers are present year-round.

Reproduction The flowers produce many small cylindrical seeds.

Propagation Cuttings or rooted nodes readily establish hardy new colonies. It prefers full sun, but does well in open shade. Periodic enthusiastic cuttings enhance the density of colonies.

Habitat & Ecological Distribution Commonly seen trailing over rocks and sand in coastal settings, this plant is tolerant of poor, dry, salty soils of most types, but is sensitive to frost. It is widely escaped and naturalized and is often the only relic of a tended yard at old house sites.

Uses

Ornamental Often used as a ground cover in sunny locations and trailing from narrow strips of soil or planters.

Medicinal The pounded leaves have been used as a poultice on sores. A tea of *Wedelia* leaves is reported to encourage passing of the placenta, to induce abortion, and to treat fevers, colds, gonorrhea and stomach and kidney problems.

Toxic It is reported that sheep have aborted their young after grazing on it.

Ecological It is an excellent groundcover for steep slopes and substantially reduces soil losses in the first year of planting without the aid of a mulch.

Notes The genus is named to honor George Wolfgang Wedel, a 17th-century professor. The species name refers to the 3 lobed leaves. Other English common names for this plant include button flower, carpet daisy and wild marigold. This plant is noted for an unusual biochemistry containing many C_{13} acetylenes with side chains of di and tri thiophenes along with thioethers. The leaves con-

tain sesquiterpene lactones, eudesmaloides, isoflavanoids, the flavonoids wedelolactone and dimethylwedolactone and other alkaloids.

MARSH FLEABANE
Pluchea odorata   *Compositae*

Geographic Distribution The genus of 40 species is found worldwide in warm areas. This species is found from the southern U.S. through the West Indies to South America.

Form A sturdy, fibrous-stemmed, deep-rooted herb to 2 m (6 ft) tall. The alternate, downy, toothed, pointed leaves are aromatic and exude a sticky resin.

Flower & Fruit Date The pinkish-purple flat-topped flower heads are numerous when in bloom, which may be any month of the year.

Reproduction The narrow black ridged seeds have a parachute of radiating bristles for wind dispersal. Vegetative spread is sometimes accomplished with rhizomes.

Propagation The seeds germinate and grow readily in moist potting soil.

Habitat & Ecological Distribution This plant is found on the borders of marshes, salt flats and other harsh dry salty sites. Individuals in colonies of this plant are frequently separated by spaces of bare ground due to the secretion of growth-inhibiting allelopathic chemicals.

Uses

Medicinal A tea prepared from the flowering leafy stems has been used as a home remedy for colds, coughs, pneumonia, tuberculosis, rheumatism, bronchitis, high blood pressure, toothache, fainting, fever and various women's complaints.

Toxic A potent, as-yet-unnamed alkaloid has been extracted from the leaves and stems.

Physical The pungent odor of the crushed leaves is reputed to serve as an insect repellent.

Ecological As a provider of shade and organic matter from shed leaves, it prepares the way for less aggressive plants to colonize disturbed sites.

Notes The genus is named after Pluche, an 18th-century French naturalist. The species name refers to the strong odor of the crushed leaves. Sour bush is another common name. The foliage and resin it exudes contain high levels of potassium nitrate, chlorogenic acid, beta-amyrin acetate, campesterol, the alkaloid betaine, the methoxyflavone artemetin, numerous eudesmane sesquiterpene derivatives, several C13 thienyl acetylenes with side chains containing chlorine and 10 different flavonols. It limits competition with other plants by producing a sesquiterpinoid dihydroxy ketone named cuauhtemone which inhibits the growth of seeds.

Wild lettuce
Lactuca intybacea　　　　*Compositae*

Geographic Distribution The genus has about 80 species worldwide. The natural range of this species is the American tropics, but it has now been introduced and naturalized in tropical Africa and Asia.

Form A herbaceous annual to 1 m (3 ft) tall with a central stalk and toothed, lobed, lettucelike leaves. Young plants will often become red-tinged due to the production of anthocyanins if exposed to the cold.

Flower & Fruit Date Up to 35 of the small yellow flowers are borne in heads throughout the year.

Reproduction The tiny elongate seeds are dispersed with white parachutes similar to dandelions.

Propagation In several species of lettuce, light is needed to promote germination. Thus the seeds should be planted on the surface of moist soil.

Habitat & Ecological Distribution This is often one of the pioneer plants growing on disturbed sites.

Uses

Ornamental This herb is so vigorous that it can be massed to provide a bed of green along upper parts of the beach or other harsh salty sites.

Medicinal A tea from the leaves has been used to treat sore throats. A sedative, lactucin, and a hypoglycemic drug, lactupicrin, have been extracted from the leaves.

Edible Young plants and leaves are good cooked, but they are bitter when raw. Goats and other livestock feed on it in the wild.

Ecological Several species of wildlife feed on the fleshy leaves.

Notes The genus name is derived from the Latin *lac* meaning "milk," referring to

the milky juice. The species name refers to the resemblance of this plant to chicory (*Cichorium intybus*). The generic name *Launaea* has also been used for this plant. Several acetylene and polyacetylene compounds, carotenes and the cinnamic acid derivatives caffeic and chlorogenic acid have been extracted from the foliage.

BEACH SUNFLOWER
Helianthus debilis   *Compositae*

Geographic Distribution The genus has at least 60 species in the New World.

Form A coarse, prostrate annual herb with rough, alternate, hairy leaves. Individual plants grow to 1.2 m (4 ft) in length but typically form larger clumped colonies seldom more than 15 cm (12 in) tall.

Flower & Fruit Date The 5-cm (2-in) bright yellow flowers with maroon centers are found throughout the year. Yellow flavanols and carotenoids make up the flower pigments. The flavanols absorb ultraviolet light and occur at the petal bases serving as nectar guides for visiting insects.

Reproduction Established clumps maintain themselves and spread effectively by self-seeding. The phytomelanines providing the dark pigment in the seeds start with secretion of colorless oily substances in intercellular spaces. These deposits gradually solidify and darken in a process resembling the polymerization of the highly unsaturated acetylenes typical of this family.

Propagation Many sunflower seeds exhibit a highly dormant phase and may require a lengthy moist period for germination. Propagation from cuttings and transplant usually works well. The beach sunflower should not be pampered. If left alone in full sun on a sandy location it will meander and form spreading clumps. Excess water and fertilizer are to be avoided, although additional water and a sparsely applied balanced fertilizer at planting aids establishment. Growth rate is rapid on ideal sites. Heavy clipping is tolerated once a stand is established.

Habitat & Ecological Distribution It is normally found thriving on dry, salty, sandy dunes and beaches. Under dry conditions growth rate is proportional to available moisture.

Uses

Ornamental This small sunflower serves as an attractive ground cover and sand binder on dunes, banks or other dry sandy sites.

Medicinal The dune sunflower is a strong allergic sensitizer due to the presence of sesquiterpene lactones. Allergic sensitivity may extend to reactions in response to other members of this family.

Edible Inulin extracted from Helianthus roots is modified into fructose by the human digestive system and is better tolerated than other carbohydrates by diabetics. It is used in the preparation of bread for diabetics.

Toxic This plant is inclined to induce allergenic responses in sensitive individuals.

Ecological This plant is effective in helping to stabilize windblown sand. The flowers are very attractive to butterflies. The characteristics which reduce transpiration while maintaining photosynthesis are being considered for genetic incorporation in the cropped forms of sunflowers.

Notes The genus name is from the Greek *helios*, meaning "sun" and *anthos*, meaning "flower." The species is Latin, meaning "crippled" or "weak" referring to the decumbent or prostrate growth habit which is in strong contrast to the vigorous upright growth habit of the cultivated sunflower. This plant is resistant to powdery mildew and attempts are being made to transfer this resistance to cultivated sunflowers. The young foliage contains the furanoheliangolide 17,18-dihydrobudlein. Typical of the sesquiterpene lactones found in this genus, this compound is antimicrobial, inhibits plant growth, and protects against insect predation. Other compounds have been found which inhibit the growth of the larvae of the sunflower moth and other lepidoptera. The seed oil contains the common fatty acids along with epoxy and hydroxy fatty acids with conjugated double bonds. Oleanic acid and some of its derivatives are the main sapogenins. The diterpenoids present such as angeloylgrandifloric acid and tetrachyrin belong primarily to the kaurene series and may be accompanied by trachylobane compounds. Amino acid sequences for the protein cytochome C have been found in the family.

154

SELECTED REFERENCES

Allen, O.N. and E.K. Allen. 1981. *The Leguminosae: A Sourcebook of Characteristics, Uses and Nodulation.* Madison, WI: University of Wisconsin Press.

Armitage, A. 1989. *Herbaceous Perennial Plants.* Athens GA: Varsity Press.

Ayensu, E.S. 1981. *Medicinal Plants of the West Indies.* MI Reference Publications.

Blackwell, W.H. 1990. *Poisonous and Medicinal Plants.* Englewood Cliffs, NJ: Prentice Hall.

Bush, C.S. and J.F. Morton. 1968. *Native Trees and Plants for Florida Landscaping.* Tallahassee, FL: Florida Department of Agriculture. Bulletin 193.

Bush, C.S. 1969. *Flowers, Shrubs, and Trees for Florida Houses.* Tallahassee, FL: Florida Department of Agriculture. Bulletin 195.

Chapman, V.J. 1976. *Mangrove Vegetation.* Vaduz, Germany: J. Cramer.

Correll, D.S. and H.B. Correll. 1982. *Flora of the Bahama Archipelago.* Vaduz, Germany: J. Cramer.

Craighead, F.C. 1971. *The Trees of South Florida.* Miami, FL: University of Miami Press.

Dawson, E. and M. Foster. 1982. *Seashore Plants of Southern California.* Berkeley, CA: University of California Press.

Duke, J.A. 1985. *CRC Handbook of Medicinal Herbs.* Boca Raton, FL: CRC Press.

Graetz, K.E. 1973. *Seacoast Plants of the Carolinas.* Raleigh, NC: U.S.D.A. Soil Conservation Service.

Hitchcock, A.S. 1936. *Manual of the Grasses of the West Indies.* Washington, D.C.: U.S.D.A. Misc. Pub. 246.

Honychurch, P.N. 1986. *Caribbean Wild Plants and Their Uses.* London: Macmillan Caribbean.

Howard, R.A. et. al. 1974-1989 (6 vols.). *Flora of the Lesser Antilles.* Jamaica Plain, MA: Arnold Arboretum, Harvard University.

Jantzen, D.H. (ed) 1983. *Costa Rican Natural History.* Chicago: University of Chicago Press.

Kinghorn, D.A. (ed) 1979. *Toxic Plants.* New York: Columbia University Press.

Kingsbury, J.M. 1964. *Poisonous Plants of the U.S. and Canada.* Englewood Cliffs, NJ: Prentice Hall.

Lampe, K.F. and M.A. McCann. 1985. *AMA Handbook of Poisonous and Injurious Plants.* Chicago: American Medical Assoc.

Leung, A.Y. 1980. *Encyclopedia of Common Natural Ingredients Used in Food, Drugs and Cosmetics.* New York: John Wiley.

Lewis, W.H. 1977. *Medical Botany.* New York: John Wiley.

Little, E.L. and F.H. Wadsworth. 1964. *Common Trees of Puerto Rico and the*

Virgin Islands. Washington, D.C.: U.S.D.A. Forest Service.

Little, E.L., R.O. Woodbury, and F.H. Wadsworth. 1974. *Trees of Puerto Rico and the Virgin Islands.* Vol. 2. Washington, D.C.: U.S.D.A. Forest Service.

Menninger, E.A. 1964. *Seaside Plants of the World.* New York: Hearthside Press.

Morton, Julia F. 1971. *Plants Poisonous to People in Florida.* Miami, FL: Hurricane House.

Morton, Julia F. 1977. *Major Medicinal Plants: Botany, Culture, and Uses.* Springfield IL: Charles Thomas.

Morton, Julia F. 1981. *Atlas of Medicinal Plants of Middle America: Bahamas to Yucatan.* Springfield, IL: Charles Thomas.

National Academy of Sciences. 1979. *Tropical Legumes: Resources for the Future.* Washington, D.C.: Author.

Nelson, Gil. 1994. *The Trees of Florida.* Sarasota, FL: Pineapple Press.

Oakes, A.J. and J.O. Butcher. 1962. *Poisonous and Injurious Plants of the U.S. Virgin Islands.* Washington D.C.: U.S.D.A. Agriculture Research Service. Misc. Pub. 882.

Oliver-Bever, B. 1986. *Medicinal Plants in Tropical West Africa.* New York: Cambridge University Press.

Record, S.J. and R.W. Hess. 1943. *Timbers of the World.* New Haven, CT: Yale University Press.

Reimold, R.J. and W. H. Queen (eds) 1974. *Ecology of Halophytes.* New York: Academic Press.

von Reis, S. and F.J. Lipp Jr. 1982. *New Plant Sources for Drugs from the New York Botanical Garden Herbarium.* Cambridge, MA: Harvard University Press.

Scurlock, J.P. 1987. *Native Trees and Shrubs of the Florida Keys.* Pittsburg, PA: Laurel Press.

Seaforth, C.E., C.D. Adams and Y. Sylvester. 1983. *A Guide to the Medicinal Plants of Trinidad and Tobago.* Port of Spain, Trinidad: Commonwealth Secretariat.

Tomlinson, P.B. 1986. *The Botany of Mangroves.* New York: Cambridge University Press.

Whistler, A. 1980. *Coastal Flowers of the Tropical Pacific: A Guide to Widespread Seashore Plants of the Pacific Islands.* Kalaheo, Kauai, HI: Pacific Tropical Botanical Garden.

Index